INTERNETWORKING LANs AND WANs
Concepts, Techniques and Methods
Second Edition

Internetworking is one of the fastest growing markets in the field of computer communications. However, the interconnection of LANs and WANs tends to cause significant technological and administrative difficulties. This updated version provides valuable guidance, enabling the reader to avoid the pitfalls and achieve successful connection.
Due 1998 0 471 97514 1

THE MULTIPLEXER REFERENCE MANUAL

Designed to provide the reader with a detailed insight into the operation, utilization and networking of six distinct types of multiplexers, this book will appeal to practising electrical, electronic and communications engineers, students in electronics, network analysts and designers.
1993 0 471 93484 4

PRACTICAL NETWORK DESIGN TECHNIQUES

Many network design problems are addressed and solved in this informative volume. Gil Held confronts a range of issues including through-put problems, line facilities, economic trade-offs and multiplexers. Readers are also shown how to determine the numbers of ports, dial-in lines and channels to install on communications equipment in order to provide a defined level of service.
1991 0 471 93007 5 (Book)
0 471 92942 5 (Disk)
0 471 92938 7 (Set)

NETWORK MANAGEMENT
Techniques, Tools and Systems

Techniques, tools and systems form the basis of network management. Exploring and evaluating these three key areas, this book shows the reader how to operate an effective network.
1992 0 471 92781 3

Please refer to the back endpapers for further details

NETWORK-BASED IMAGES

NETWORK-BASED IMAGES
A Practical Guide to Acquisition, Storage, Conversion, Compression and Transmission

Gilbert Held
4-Degree Consulting
Macon, Georgia,
USA

JOHN WILEY & SONS
Chichester • New York • Weinheim • Brisbane • Singapore • Toronto

Copyright © 1997 by John Wiley & Sons Ltd,
Baffins Lane, Chichester,
West Sussex PO19 1UD, England

National   01243 779777
International (+44) 1243 779777
e-mail (for orders and customer service enquiries): cs-books@wiley.co.uk

Visit our Home Page on http://www.wiley.co.uk or http://www.wiley.com

Other Wiley Editorial Offices

John Wiley & Sons, Inc., 605 Third Avenue,
New York, NY 10158-0012, USA

Wiley-VCH Verlag GmbH,
Pappelallee 3, D-69469 Weinheim, Germany

Jacaranda Wiley Ltd, 33 Park Road, Milton,
Queensland 4064, Australia

John Wiley & Sons (Asia) Pte Ltd, 2 Clementi Loop #02-01,
Jin Xing Distripark, Singapore 129809

John Wiley & Sons (Canada) Ltd, 22 Worcester Road,
Rexdale, Ontario M9W 1L1, Canada

British Library Cataloguing in Publication Data

A catalogue record for this book is available from the British Library

ISBN 0 471 97357 2

Typeset in $10\frac{1}{2}$/12pt Bookman by Aarontype Ltd, Bristol
Printed and bound in Great Britain by Bookcraft (Bath) Ltd
This book is printed on acid-free paper responsibly manufactured from sustainable forestry, for which at least two trees are planted for each one used for paper production.

CONTENTS

PREFACE

The old adage 'one picture is worth a thousand words' has finally been accepted in the world of personal computing. Although the first decade of personal computing was perhaps conspicuous by the absence of image-based applications, this has rapidly changed. In fact, the number of image-based applications has probably increased at a rate equal to or exceeding the budget deficits of many countries. Applications that were a figment of our imagination a few years ago today involve the transmission of millions of images every day. Telemedicine, real estate, personnel databases, automobile fender benders, and of course the World Wide Web all depend on the transmission of images.

Although the transmission of images is visually appealing and it enhances the conveyance of information, it is not without cost. For the network manager and administrator, that cost is primarily related to the high consumption of bandwidth by images, usually resulting in costly network upgrades to alleviate the adverse effect of the transmission of images on other network applications. Those upgrades can include an expansion of disk storage, as well as the modification or replacement of an existing transmission infrastructure. For the individual computer user, the increase in the use of image-based applications can result in extended transmission times and can rapidly reduce available disk storage applications. In addition, because the cost of the public switched telephone network and packet networks are proportional to transmission time, the conveyance of information in the form of images can adversely affect communications bills. Thus, both individual PC users, as well as network managers and administrators, have a vested interest in learning how to effectively and efficiently work with network-based images, the focus of this book.

This book was written to provide both individual PC users and network managers and administrators with detailed practical information concerning the acquisition, storage, conversion, compression and transmission of images. By understanding how to effectively and efficiently work with images you may be able to avoid or alleviate costly network upgrades, reduce transmission times and enhance user productivity.

The information presented in this book is based on actual experience obtained from managing a complex, nationwide data communications network that connects users on over 100 LANs throughout the United States to seven mainframe computers as well as to hundreds of network servers. By having experienced the problems associated with incorporating images into network-based applications as well as having developed solutions to those problems, I have learned through attendance at the 'school of hard knocks' valuable information that will hopefully allow you to save days and weeks of effort, as well as providing verified techniques that can be used to conserve bandwidth and reduce both data storage and data transmission time. To paraphrase a great leader of the century, 'We have nothing to fear from network-based images if we understand how to effectively work with those images.'

As an author who highly values reader comments, I encourage you to contact me with your suggestions. No person has a full grasp on all aspects of technology and I certainly welcome your input concerning topics that you might like to see in a new edition of this book. You can write to me in care of my publisher.

Gilbert Held
Macon, GA

ACKNOWLEDGMENTS

The preparation of a book is a team effort, requiring the work of many persons in addition to the author. Thus, I would be remiss if I failed to acknowlede the fine work of Mrs. Linda Hayes in converting this author's handwritten notes and drawings into a manuscript; Ms. Ann-Marie Halligan who as Publisher at John Wiley & Sons backed this project; and Mr. Ian Stoneham who guided the manuscript through the production process. Special thanks goes to Jin-Sinh Ho for permission to include his DISPLAY program in the book, to Chris Komnick of Group 42 to include their WebImage program, to Chris Anderson of JASC Inc. for permission to include their comprehensive image manipulation program Paint Shop Pro, to Laura Shook of Cerious Software Inc. for permission to include their ThumbsPlus v3.0e and thumbsPlus v2.03 programs, and to Richard Marks for his UUENCODE/ UUDECODE program pair, all included on the CD-ROM accompanying this book. I would also be remiss if I did not thank my family for their cooperation and understanding during the long evenings and weekends that I hibernated in my office drafting the book you are now reading.

1

INTRODUCTION

This book was written as a comprehensive guide to the acquisition, storage, conversion, compression and transmission of images. The goal of this book is to provide network managers and administrators, as well as Web Masters, Web Mistresses and individual personal computer users, with detailed information that illustrates techniques to effectively and efficiently incorporate images into a variety of network-based applications. To accomplish this goal, this book assumes that readers have no prior knowledge of the technical details associated with computer-based images. Thus, this book also serves as a basic guide or tutorial concerning computer-based imaging, as well as its intended focus, which is to provide readers with detailed information that illustrates how to effectively and efficiently use network-based images.

As an introductory chapter, we will first focus our attention on the rationale for being concerned about network-based images. In doing so, we will examine a variety of image-based applications and their potential effect on both individual computers and networks. As this is an introductory chapter, our discussion concerning the effect of images on computers and networks will result in the introduction of some basic image-related terms; however, a much more comprehensive description of images will be left for the next chapter in this book. Once we have an appreciation for the rationale for being concerned about the appropriate use of images, we will turn our attention to a preview of the other chapters in this book. In doing so, we will discuss the content and focus of each chapter.

In addition to providing readers with a synopsis of the material to be presented in succeeding chapters, this preview, as well as the table of contents and index, will enable readers to directly locate information of immediate interest.

1.1 A TOUR OF IMAGE APPLICATIONS

The rationale for being concerned about network-based images is really quite simple; it is the applications. Until a few years ago image-based applications were an exception to the preponderance of text-based applications. Since the early 1990s the number of image-based applications has experienced a quantum leap, with images now representing the rule rather than the exception when we discuss application programs and their operation and utilization. To obtain an appreciation of the scope and depth of the use of images in applications, consider Table 1.1 which lists potential network-based applications that could occur in seven diverse functional areas within a corporation. By briefly discussing each application, we can note how pervasive the transmission of images on both an intra- and internet basis has become, and we shall become familiar with some of the network performance issues associated with the storage and transmission of images.

Catalog sales

Although most, if not all, readers are familiar with one or more mail order catalogues that commonly clutter our mail boxes, if you take your exotic car into a garage, need a part for your Hungarian watch, or require the replacement of the trigger on

Table 1.1 Examples of image based applications

Catalog sales
   visual image of products

Computer-based training
   screen displays

Insurance
   photographs of accidents

Personnel database
   employee verification

Real estate
   home interiors and exteriors

Telemedicine
   radiological results

World Wide Web
   pictures of persons, places and things

your old Smith & Wesson, the repair person will more than likely call another type of catalog sales. At the called location an employee may use a computer to access a database that contains text descriptions as well as images of parts, including the dimensions of certain parts and their component numbers, if the part represents an assembly consisting of multiple parts. By conversing with the customer and obtaining the ability to have a visual image of the part displayed, employees are better able to answer customer queries, boost their productivity and lower the percentage of returned parts. In many organizations such databases now reside on local area network (LAN) servers, with employees accessing centralized catalog files to answer customer queries.

As most organizations have a finite budget, the method used for the storage of images can have a direct effect on the bottom line. If images are stored as they are, without compression, and they represent high-resolution photographs with intricate details and lots of color, each image may use a significant portion of a server's disk storage capacity. As the image is transmitted to client workstations, its transmission can adversely affect other network users. Thus, the storage of images can represent trade-offs concerning the use of data compression to reduce data storage and data transmission.

Computer-based training

Computer-based training (CBT) represents a popular and cost-effective mechanism to train employees in different subjects. As you might expect, some of the more popular CBT courses are in computing, focused on the use of different windows-based operating systems. Although you can purchase single-use CBT training courses, many organizations now purchase a site license and place the CBT course on one or more LAN servers. Needless to say, as employees use one or more CBT courses, chances are very high that images will be transferred from the server to the employee's PC, representing all or portions of windows-based operating systems which the course uses to illustrate different concepts. Often a group of employees working on a new CBT course can bring a network to its knees as their training sessions result in a large amount of bandwidth-intensive traffic flowing on the LAN.

Sometimes the manner by which applications use images cannot be improved through the use of software. This is especially true when you purchase some CBT training courses that integrate text and graphics in a proprietary manner. In such situations you may have to consider other options, such as restructuring a network, migrating CBT courses to a separate server, or using LAN switches to enhance access to image-based applications.

Insurance

In the past, after a fender-bender and a call to your insurance agency, a representative would arrive at your home or office, use a Polaroid camera to take pictures of the damage, fill out a few forms, and mail everything to the home office. At the mercy of the post office, you might receive a check several weeks later. Today, many insurance companies equip their field representatives with laptop computers, cellular modems and digital cameras. After taking a digital photograph of the damage and filling out a form on the laptop, the representative pops a PC card out of the camera and inserts it into his or her notebook. After selecting an appropriate communications program, the contents of the form and the photograph or series of photographs recorded by the digital camera onto the PC Card are transmitted via the cellular modem to the home office. As the representative drives away to his next appointment, employees at the home office begin to process your insurance claim.

Although digital cameras are increasingly being recognized as a mechanism to boost employee productivity, their resulting images require careful consideration. As cellular communications are billed by usage, the use of a resolution beyond that necessary to show appropriate damage results in additional storage that requires additional transmission time. This not only adds to the cost associated with field representatives transmitting accident reports to the home office, but, in addition, increases the data storage requirements of the home office which, due to state regulations, they may have to maintain for many years. Once again, it is important to fully understand the different aspects of an image-based application.

Personnel data base

It is your first day of work and during induction your photograph is taken for an identification card as well as for use by the personnel office. That office may scan your photograph and enter it into a visual database, containing such information as your date of hire, education, skills, the department or organization that you are assigned to, security clearance, and similar personnel information. A few weeks after you are hired you are requested to visit a branch office to obtain information about a special project. Although you have an identification card, your organization is very security-conscious. After all, it is relatively easy to super-impose a new photograph on an ID and re-laminate it. When you arrive at the distant location and check in at the guard station, the guard accesses the personnel database and obtains permis-sion to retrieve your photograph and vital statistics concerning your age, weight and height. After glancing at you, your ID, and the console, the guard gives you a temporary building pass and calls the party you are visiting to escort you from the lobby.

In this imaging application, the image might be transmitted from one LAN onto another via the use of a wide area network (WAN) that connects the two distant locations. Thus, the trans-mission of the image could affect operations on multiple local area networks as well as traffic competing for service via the wide area network.

This description of an application illustrates an important concept concerning network-based images. That is, the trans-mission of images can have a significant effect on different organizational locations, and the appropriate balance between transmission response time, image resolution and the applica-tion must be carefully considered.

Real estate

In many areas it is quite a familiar site to see an estate agent stop-ping a car at an intersection or in front of a home on a Sunday morning, taking out an Open House sign and pressing it into a grassy area. Although not quite as common, but rapidly increas-ing in use, is a digital camera which is becoming part of the 'tools of the trade'. Today many agents are using digital cameras to take photographs of exteriors and interiors of properties to assist in their marketing effort. On returning to their office, the digital images are downloaded from the camera into visual databases maintained on a LAN, or are perhaps placed on a World Wide Web server along with a description of the property.

When potential clients visit the real estate office, they can work with an agent to define the type of property they are seeking with respect to price range, type of home, location, school district and other parameters. Using that information, the agent can query the visual database and display photographs of the interiors and exteriors of different properties, enabling the prospective client to narrow the search before leaving the office for a visual inspection of properties. Not only does this save wear and tear on the agent's car, but in addition it saves that precious commodity known as time and allows the agent to service additional clients.

Currently, the use of digital cameras by estate agents is essentially limited to individual firms and is not part of a multiple listing service. This means that photographs taken through the use of digital cameras are primarily placed on individual LANs or Web servers. However, due to the significant advantages associated with the incorporation of images into real estate listings, it is probably only a matter of time until images are added to multiple listing databases. At that time, the manner in which images are placed into the database, including their resolution, use of color and image size in pixels, will be extremely important considerations owing to the use of the relatively slow public switched telephone network to access centralized multiple listing databases. This is because it would be impractical to expect agents and their potential customers to wait five or ten minutes to view one or a few photographs of individual properties when they are searching through a potential series of 10 to 20 or more residences. For this type of environment the ability to compress images and reduce their transmission time becomes extremely important, even if the cost of a local telephone call is not affected by a reduction in transmission time.

Telemedicine

You are probably aware of the problems that small communities in rural areas are experiencing in staffing hospitals and clinics. You are probably also aware of some of the methods being used to provide medical support to those communities. Among those methods is one commonly referred to as telemedicine, meaning physician assistance provided via communications.

One of the features associated with telemedicine is the transmission of radiological images, such as CAT and MRI scans as well as conventional x-rays. The transmission of such images enables members of the medical profession in one location to provide guidance and assistance to physicians in another location.

Unlike conventional images in which the loss of minor details is normally acceptable, the loss of even one pixel on a radiological image is normally not acceptable. This means that any improvement in the transmission and storage of such images obtained through the use of data compression is restricted to the use of lossless data compression algorithms. Here the term lossless refers to the fact that the data decompression process results in the exact duplication of the original image on a pixel-by-pixel basis prior to the image being compressed. In comparison, a lossy compression method results in the loss of pixels when the compressed image is decompressed. In general, lossy compression results in a larger data reduction than lossless compression, and the fact that some lossy compression methods enable you to specify the degree of loss. Thus, the processing of radiological images with respect to making them more efficient for storage and transmission is more restrictive than other types of images, and it deserves careful consideration by network managers, administrators and computer users.

World Wide Web

From a mere concept in 1989, by 1997 the use of the World Wide Web had literally exploded. Today, many advertisers end their television commercials with their Web addresses. Such addresses

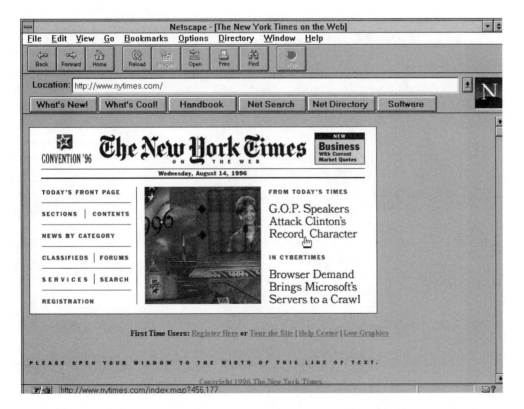

Figure 1.1 Home page of the *New York Times* that uses a large map-based image for user selection

also commonly appear in newspapers, magazines and employee business cards. Accompanying the growth in the use of the Web is the realization by business, government and academia that hosting a Web site can be an expensive proposition. In addition, a poorly constructed site can result in potential customers literally 'twiddling their thumbs' waiting to retrieve information, increasing the risk that they can either receive a timeout message or decide to click on their browser's stop button and select a new address to visit.

In concluding this section, we will examine some popular Web pages and discuss the reason why some pages take considerably more time to load than other pages, illustrating another network-related image issue that you must consider.

Figure 1.1 illustrates the World Wide Web home page of the *New York Times* on Wednesday, August 14, 1996. This page contains one image in the form of a map, which enables users to move their cursor over different portions of the map and click on a text entry to jump to a particular area. Figure 1.2 contains the source code

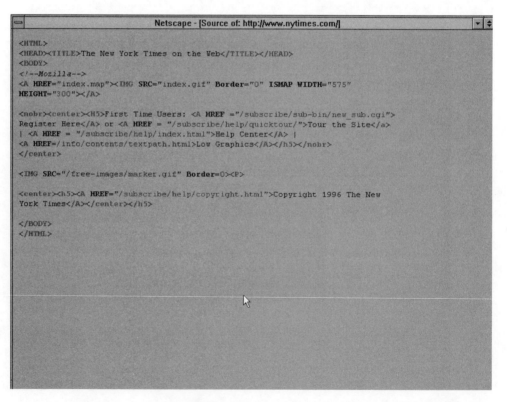

Figure 1.2 HyperText Markup Language statements used to create the Web page shown in Figure 1.1

used to generate the home page illustrated in Figure 1.1. That code is in the form of HyperText Markup Language (HTML) statements. Although we will discuss the use of image-related statements later in this book, if you focus your attention on certain HTML statements in Figure 1.2 you can determine information about the image. For example, from that listing you can ascertain that the image was sorted in a GIF file format and is relatively large, as it has a width of 575 pixels and height of 300 pixels. This means that without considering the color of the image, 575×300 or 172 500 pixels or over 21 000 bytes must be downloaded to view the image. In comparison, a full screen of text represents 80 columns by 25 lines or a total of 2000 bytes. Thus, without considering the color details of the map image, it requires over ten times the transmission time of a full screen of text. According to tests performed by Netscape Communications Corporation, a total of 6.4 seconds were required to load this page into Netscape's Navigator 3.0 browser. For comparison purposes, let us view the home page of Silicon Graphics, illustrated in Figure 1.3.

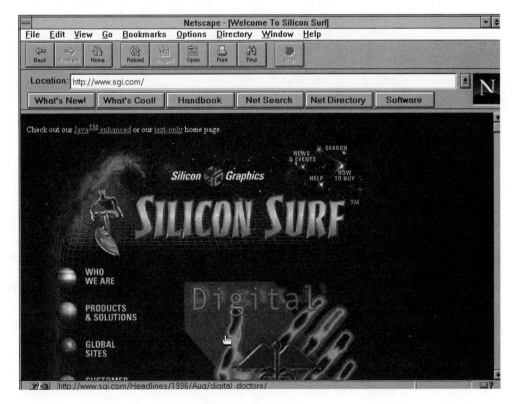

Figure 1.3   The Web home page of Silicon Graphics

You will agree that this Web page is much more interesting than the previous page illustrated in Figure 1.1, yet the time required to load this page into Netscape's Navigator 3.0 browser is only one-tenth of a second longer. The primary difference between the two home pages can be attributed to the use of graphics and screen background, items that we will discuss in later chapters.

As indicated in this section, the use of computer-based images is pervasive. In addition, there are many issues associated with the storage and transmission of images that govern their efficient and effective use, as well as affecting your bottom line. Thus, it is important to understand how images are acquired, stored, converted, compressed and transmitted because this knowledge will provide us with the ability to recognize tools and techniques that we can employ to more efficiently and effectively use images. With this in mind, let us preview the other chapters in this book.

1.2 BOOK PREVIEW

As previously mentioned, this book was written using the assumption that many persons will have either a limited knowledge or a lack of knowledge concerning the technical details associated with images. As knowledge of such details is essential to understand how different techniques can be employed to more effectively and efficiently work with images, beginning in Chapter 2 we will focus our attention on the different types of images and the technical details associated with different types of image.

Image fundamentals

In Chapter 2 we turn our attention to obtaining a detailed under-standing of the different types of images, how they are repre-sented, and the effect of color on images. Using this information as a base, we will then examine the data storage requirements of different types of images to obtain the ability to examine the effect that results from altering images to enhance their transmission and data storage.

Image formats

Continuing our technical exploration of images, in Chapter 3 we will turn our attention to the characteristics of several popular image file formats. In this chapter we will note their support of data compression and its effect on the transmission and storage of images. Through the use of an image that will be converted into a variety of image formats, we will obtain an appreciation of the effect of acquiring images using different image formats or converting an image from one format to another.

Acquisition techniques

The primary focus of Chapter 4 is on how images are input into a computer for manipulation. In this chapter we will examine the acquisition time, storage and other parameters associated with the use of different acquisition techniques as well as their effect on the subsequent use of the image in different applications.

Images and the LAN

In Chapter 5 we turn our attention to examining the effect of the placement of images on local area networks. In this chapter we will note the effect of several images on transmission delays, and discuss problems associated with multicast image-based applications.

Once we have obtained an appreciation for the technical details associated with images and the potential problems they can cause, we can focus our attention on enhancing the use of images on networks. In the second part of Chapter 5 we will examine the application of a variety of hardware and software solutions to different image problems that occur or that can be expected to occur on local area networks.

Integrating images on Web pages

No book on images would be complete without a detailed discussion of the use of images in World Wide Web (WWW) documents. In Chapter 6 we turn our attention to the use of HTML statements to place images on Web pages, how to display different types of images, and the effect obtained from the use of different types of images. In doing so we will create several examples of Web pages and compare and contrast the efficiency of each page with respect to transmission time and storage.

Determining the Internet Web connection

One of the key problems facing organizations that want a presence on the Web is in determining an appropriate Internet connection. If the connection is too slow, you may lose viewers and potential customers or clients for your products. If the connection is beyond that necessary to support your site, you are probably paying too much for an unnecessary level of service. In Chapter 7 we turn our attention to this problem by obtaining a methodology for computing an appropriate operating rate to connect a Web server to the Internet.

Image transmission techniques

Assume that you have just located a document produced by a word-processing program that contains an image you want to use on your organization's Web server. How do you transfer it? What happens if a branch office requests you to transmit several pages of a manual containing images and your e-mail system is limited to supporting 7-bit ASCII? These are two of several problems that are solved in Chapter 8, where we examine image transmission techniques.

Shareware operation

In the concluding chapter of this book we will examine the operation and use of several freeware, shareware and tradeware programs on the CD-ROM accompanying this book. Such programs were selected for inclusion in this book, as they provide users with the ability to manipulate images such that the resulting manipulation can provide more efficient and effective means to store and transmit such images.

IMAGE FUNDAMENTALS

In this chapter we turn our attention to the basic technical details associated with the representation of images. To do so we will first discuss the two primary types of image as well as focusing our attention on how those images are represented. Concerning the latter, we will examine how black and white, gray scale, and different types of color on an image are represented in terms of bits required to store information concerning the color of each point on an image. This information will be used to compute the data storage requirement associated with different types of images which is proportional to their transmission time. As we build on our knowledge of images we will become acquainted with a variety of image related terms that will be used in subsequent chapters in this book.

2.1 IMAGE CATEGORIES

Images can be categorized by the manner by which they are stored, displayed and manipulated. There are two major categories or types of images, raster and vector.

Raster images

A raster image, perhaps more correctly termed an image stored in a raster format, is represented by a series of picture elements or pixels of equal size. The raster format breaks an image into a grid or matrix of pixels and records color information for each pixel. If we consider the basic raster image without color, other than black and white, the term bitmap is used to describe the shapes in the image as a pattern of dots or pixels. Although the term bitmap actually describes a bi-level black-and-white raster image, it is also commonly used (although technically incorrect) to refer to color raster images. To be technically accurate we will use the term bitmap to refer to a bi-level black and white raster image, and the term raster to refer to any image formed by a grid of pixels, including a bitmap.

Resolution

Although many of us are familiar with the term resolution when it is applied to a monitor or display, this term is also applicable to the acquisition, storage and printing of an image. When we discuss the resolution of an image, we are referring to three dimensions of an image: its width, height and color depth. For example, a commonly used VGA display mode is 640 pixels wide by 480 pixels high, with eight bits per pixel used to display one of 256 colors from a pallette of 256 colors for each pixel. Here the use of eight bits per pixel refer to the color depth of the display. The color depth, which is also referred to as bits per pixel or bit depth, defines the number of colors that can be assigned to the pixel and is an important consideration when working with images.

Color depth

A raster image with a color depth of one bit per pixel is restricted to providing a black and white representation, as only two choices (0 and 1) are available per bit position. Most raster image formats support more than one bit per pixel, permitting more than one level of color per image. Table 2.1 lists some common bits-per-pixel values supported by popular raster image formats and the corresponding maximum number of colors. In chapter 3 we will discuss and describe actual raster file formats.

In examining Table 2.1, several items are worth noting, as they may influence the manner by which you use images. First, the capability of many personal computers, including most PCs manufactured before 1994, are limited to displaying a maximum of 256 colors. Secondly, a color depth of 24 bits is commonly referred to as 'true color', as human eyesight cannot normally distinguish colors beyond those supported by a 24-bit color depth. Although a few scanners and display adapters and monitors can now support a color depth of 32 bits per pixel, most persons will fail to see any difference between the use of that colour depth and a 24-bit color depth.

Table 2.1 Maximum color support versus bits per pixel

Bits per pixel	Maximum number of colors
1	2
2	4
4	16
8	256
16	32 768
24	16 777 216

Basic computations

Knowledge concerning the resolution of a raster image is important for several reasons. First, if you know the pixel dimensions and color depth you can compute the number of bytes required to store the image. Although this computation does not include the header of the file, which may use a number of bytes to indicate certain file format information it normally provides a very accurate file storage size. Thus, the uncompressed size of a raster image file can be expressed in bytes as follows:

$$\frac{\text{pixel width} \times \text{pixel height} \times \text{color depth} + \text{header bytes}}{8 \, \text{bits/byte}}$$

For example, consider an image of 640 by 480 pixels with a color depth of 24 bits/pixel. Then the file size in bytes required to store this image without considering file header information becomes

$$\frac{640 \times 480 \times 24}{8} = 921\,600 \, \text{bytes}$$

A second reason to know the resolution of a raster image is to detemine if an image can cover a predefined area. Often, the resolution of scanners, monitors and other products that acquire or display images is expressed in dots per inch (dpi), which is a synonym for pixels per inch (ppi). The relationship of dpi to pixels is given by the following formula:

$$\text{pixels} = \text{dpi} \times \text{inches}$$

Thus, the resolution expressed in dpi or ppi multiplied by the length in inches equals the pixel length of an image. For example, assume that you are using a scanner set to a resolution of 300 dpi to convert a 3×5 inch photograph into a raster file. Then the size of the scanned image would be 3×300, or 900, pixels in height by

5×300, or 1500, pixels in width. If the image was scanned using a color depth of 24 bits, then the scanned file would require

$$\frac{900 \times 1500 \times 24 \text{ pixels}}{8 \text{ pixels/byte}} = 4\,050\,000 \text{ bytes}$$

It should be noted that, although dpi and ppi are synonymous, the former term is primarily used as a measurement of the capability of an input or output device. In comparison, pixels per inch is primarily used to note the actual resolution of a digital image. A third term that requires a degree of elaboration is lines per inch (lpi), which was originally used as a printing designation to describe the density or frequency of halftone dots used to reproduce images on a printing press. Here the term halftone represents the scaling of black dots to fool the eye into seeing gray or the mixing of the four color inks (CMYK: cyan, magenta, yellow and black) such that a person thinks they are viewing a continuous-tone colour photograph. Through the use of half-toning, spots are smaller in light areas and larger in dark areas, enabling printing presses, laser printers and other devices that cannot print gray to use black ink or toner to provide the impression of gray.

The size of halftone dots is commonly so small that they cannot be viewed without magnification. Newspapers commonly use 85 lpi, and many magazines use a 133 or 150 lpi resolution. Similarly to dpi and ppi, the higher the lpi the finer the halftone resolution. For example, high-quality printing on coated paper commonly use 150, 175 and 200 lpi resolution.

Using an image

To provide an illustration of the storage associated with different types of images requires one or more images to work with. Rather than dealing with the common delays associated with attempting to request permission to use copyrighted photographs, I turned to my trusted companion who I had to bribe with a few dog treats to stand still long enough to take his picture. Once it had been developed this author used his scanner to input the digitized image of his dog named Gizmo into my computer.

Figure 2.1 illustrates the captured screen image of the operation of the Art-Scan program used with this author's scanner. In this example the scan mode was set to gray scale, which results in 8 bits per pixel being used to represent up to 256 shades of gray for each pixel.

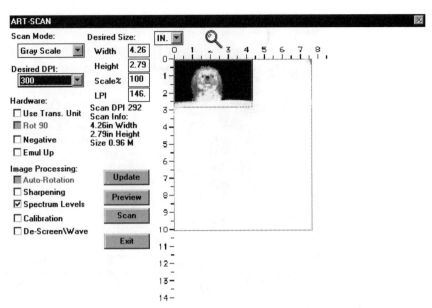

Figure 2.1   Using the program Art-Scan to scan a photograph

In examining Figure 2.1 you should note that instead of scanning the entire flatbed of the scanner or the 3×5 inch dimension of the photograph, this author reduced the desired scan area. That reduction resulted in the area to be scanned representing a width of 4.26 inches and a height of 2.79 inches, for a total of 11.8854 square inches. Based on selecting a desired dpi 300 or 90 000 pixels per square inch, you would expect the size of the image to be $90\,000 \times 11.8854$ or $1\,069\,686$ bits without considering the color depth of the image. As the selected scan mode used 8 bits or one byte per pixel for the color depth, you would expect the resulting file to require $1\,069\,686$ bytes of storage plus a few additional bytes for the file header. In actuality, the resulting file is almost 64 000 bytes larger, as the actual scan was significantly longer and slightly shorter than the desired size shown in Figure 2.1. To determine the actual dimensions of the resulting scanned image this author used the program Collage Image Manager. Figure 2.2 illustrates the display of a portion of the scanned file superimposed by the display of a Image Information window obtained by selecting the program's Image Information entry from its Window menu. As indicated by the Image Information display, the scanned image is 1460×776 pixels using color 'type' of 8 bits. Thus

$$\frac{1460 \times 776 \times 8\,\text{bits/pixel}}{8\,\text{bits/byte}}$$

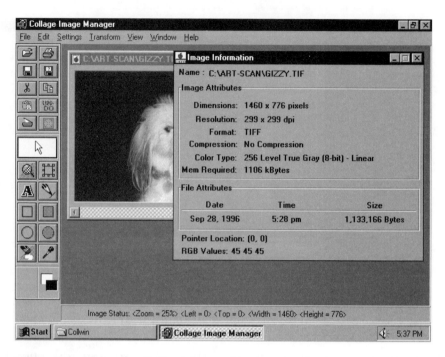

Figure 2.2 Using the Collage Image Manager to display the details of a previously scanned image

results in a data storage of 1 132 960 bytes. Note that the size entry under 'File Attributes' in Figure 2.2 indicates 1 133 166 bytes, or an additional 206 bytes. Those additional bytes are required to define the dimensions, resolution, compression and color type stored in the header of the image file which is in what is referred to as a TIF format. In Chapter 3 we will examine image file formats in detail.

As it was difficult to see Gizmo in the previous illustrations, he asked this author to show readers his obedience and dedication to photography. Deciding to accomplish two things with one illustration, because we will use his raster image as a mechanism for comparing it to a vector image later in this chapter, this author consented to his bark. Thus, Figure 2.3 illustrates this author's dog during an obedient moment.

Vector images

A second type or category by which images are displayed and stored is based on the use of mathematical algorithms that

Figure 2.3 The author's dog Gizmo in an obedient moment produced using a 256 level gray scale and resolution of 300 dpi

generate direct line segments in place of individual pixels. The resulting shape generated by a mathematical formula is referred to as a vector image.

Examples

Examples of vector images include images created using a computer-aided design (CAD) program as well as hardcopy generated by PostScript, a language which describes how graphics are displayed. As a series of mathematical algorithms are used to describe the components of an image, the storage of those algorithms requires far less storage space than storing individual pixels. This results in vector images that require considerably less storage than an equivalent raster image. However, before converting raster images into vector images, it is important to note that certain types of image are unsuitable for representation by a series of mathematical algorithms. To illustrate this, let us again use this author's dog Gizmo.

Figure 2.4 illustrates the conversion of the raster image previously shown in Figure 2.3 into a vector image. Note the loss of detail in Figure 2.4 because the computer has to interpret the pattern of pixels in the raster image as shapes, and then attempts to generate the appropriate algorithms to draw the shape.

In general, drawing programs such as Auto CAD, McDraw, etc., generate vector images. In comparison, photographs taken with a digital camera, creative art drawing using a pixel-based 'paint' program and most scanned photographs are examples of raster images. Since employee pictures in personnel files, real estate applications which include photographs of interiors and exteriors

Figure 2.4   A vector image of the author's dog which produces a considerably smaller file but loses a lot of detail

of homes and most World Wide Web pages are based on the use of raster images, the primary focus of this book will be on this category of graphics representation. In addition, because the data storage requirements of raster-based images can exceed by several orders of magnitude the data storage requirements of vector-based images, it is the former type of image that can be expected to consume significant communications bandwidth. Thus, our primary focus on raster-based images will provide an understanding of where the majority of bandwidth-associated imaging problems arise, as well as enabling a description of the methods and techniques that you can consider to alleviate those problems.

Other categories

When discussing image categories there are several additional terms that we should note. Those terms include metafiles and page description languages (PDLs). A metafile represents a file format that can contain either raster or vector graphics data. A PDL is used to describe the layout or composition of a printed page of text and graphics, which can include either or both types of images.

2.2 TERMS TO NOTE

In concluding this chapter, we will turn our attention to obtainig an overview of a few image-related terms that we will refer to later in this book. Such terms govern the proportion of the dimensions of an image, the difference between interpolation and true resolution, and the two basic types of compression that can be used to reduce the storage and transmission time of images.

Aspect ratio

The aspect ratio represents the proportion of the dimensions of an image in the form of height-to-width. For example, consider the commonly used $8\frac{1}{2} \times 11$ inch page. That page has an aspect ratio of $11:8.5$ inches or $22:17$. Often the aspect ratio is expressed as a single number by dividing the larger number by the smaller number. In the previous example, the aspect ratio could be expressed as 1.294 (22/17).

Interpolation

Scanners and video capture hardware, as well as image manipulation software, commonly provide the ability to produce interpolated resolutions up to 2400 dpi. When interpolation is selected, the values of adjacent pixels are used to create new pixels. In comparison, true resolution resulting from the capture of an image provides more shapely and minute details than obtainable from interpolation.

Due to the prohibitive cost or unavailability of image acquisition products to support a very high level of true resolution, many products include software which provides an interpolation capability to boost the resolution of a previously acquired image. Care should be taken when using interpolation, as its use can significantly boost the storage requirements of an image as well as its transmission time. To illustrate this fact, consider an image scanned using a resolution of 300 dpi that is interpolated to a resolution of 2400 dpi. At 300 dpi each square inch of the scanned image requires 300×300 or 90 000 pixels without considering its color depth. In comparison, when the image is interpolated using a resolution of 2400 dpi, the data storage requirement increases to 5 760 000 pixels per square inch, an increase by a factor of 64! This increase results from a 2400 dpi interpolation increasing the vertical and horizontal number of pixels by a factor of eight in each direction beyond that obtained from the use of a 300 dpi resolution. Thus, 8×8 is 64, which represents the storage increase resulting from the use of 2400 dpi instead of 300 dpi.

Need for compression

When working with high-resolution images it is important to use a file format that supports compression. Otherwise the size and transmission time required to move the image to another location may become unmanageable. To illustrate the need for compression, consider a 3×5 inch photograph which has a total of 15 square inches. If the image were to be stored using a 2400 dpi interpolation, a black and white photograph would require $5\,760\,000 \times 15/8$ or 10.8 Mbytes. If the interpolated image had a color depth of 8 bits per pixel the data storage required for the image would increase to 86.4 Mbytes! Even though you can now purchase a 1 Gbyte hard disk for a few hundred dollars, without compression that disk would be filled by the storage of 11 images. In addition, assume that a manager at another office calls you and requests a copy of the image. If you were limited to using a modem with a transmission rate of 28.8 Kbps, the transmission of an 86.4 Mbyte file would require

$$\frac{86.4 \,\text{Mbytes} \times 8 \,\text{bits/byte}}{28.8 \,\text{Kbps}} = 24\,000 \text{ seconds}$$

which represents approximately 6.67 hours. Clearly this would be an unacceptable amount of time, and it indicates why the use of data compression is a popular technique for reducing the data storage and transmission time of images.

Types of compression

As briefly mentioned in Chapter 1, there are two methods of data compression: lossless and lossy. Both methods of data compression use the data redundancy in a file to reduce the storage of the file. In a lossless compression method the algorithm or algorithms used to eliminate or reduce data redundancy are fully reversible, enabling the image to be restored on a pixel-by-pixel basis. In comparison, lossy compression considers blocks of pixels to be equal even if they differ by one or a few pixels. Although lossy compression can provide a much higher reduction of an image's storage requirements than the use of a lossless compression method, the file resulting from the use of a lossy compression method cannot be restored to its original state. This means that some detail of an image will be lost through the use of lossy compression.

Lossy compression is commonly used to transmit newspaper and magazine images. In comparison, lossless compression is commonly used in telemedicine applications where each pixel can be very important. In Chapter 3 we will focus our attention on the effect of the two methods of data compression as we examine different image file formats.

3

IMAGE FILE FORMATS

In this chapter we turn our attention to obtaining an overview of a relatively large number of image file formats that have been developed to store images. In doing so, we will discuss the primary use of each file format and their use of compression, if applicable. Although you may be perplexed as to why we will examine image file formats before examining image acquisition techniques, there is actually a good reason for what may appear at first glance to be a reversal of chapter ordering. By understanding the advantages and disadvantages of different image file formats in terms of resolution, color depth and storage requirements we can then give a solid foundation for discussing the various storage options provided by different image acquisition techniques, which is the subject of the next chapter.

3.1 RASTER IMAGE FORMAT BASICS

As previously noted in Chapter 2, a raster format represents a grid consisting of equally-sized pixels that are stored with color information. In addition to supporting black and white (one color bit per pixel), most raster formats support more than one level of color depth. This means that if you have the ability to select the color depth of an image you can also control to a degree its data storage requirements. Another option that is associated with many but not all image file formats is the ability to enable or disable compression, whereas a few file formats support several types of compression. Thus, the ability to vary the color depth and the use of data compression can significantly alter the data storage requirements of images. As transmission time is proportional to the storage requirements of a file, the ability to alter color depth and enable or disable data compression can significantly affect the time required to move an image via a transmission facility.

Since images were first digitized and stored as files, over 60 raster file formats have been developed to standardize their recording and viewing. Table 3.1 provides a summary of 40 raster file formats, their commonly used file extensions, their primary format source, and the support of different color depths for those file formats. Concerning the file format source, that column in Table 3.1 indicates the primary application that generates the indicated file format. Readers should note that many programs include the ability to store images in a half dozen or more file formats, whereas some programs that were developed specifically to display and manipulate images may support 30 or more image file formats.

Common formats

The importance of one or more of the raster image formats listed in Table 3.1 depend on the application you are using and its ability to support different formats. If you primarily use an application that generates raster images in one or a few formats, then those formats will be important considerations. Although this means that the importance of a particular format can and will vary between individuals, there is a core set of raster image formats that probably represent a large majority of the methods by which images are displayed and stored. Table 3.2 includes a list of ten popular file formats and a more detailed explanation of each file format than contained in Table 3.1. In the remainder of this section we will discuss nine of the file formats contained in Table 3.2. in some detail, due to their popular support by many image applications.

ART

Software Publishing Corporation was one of the first vendors to incorporate clip art images into an application. The firm's First Publisher application uses ART files that are limited to black and white images used as clip art by the application. As ART files are always black and white, the color depth of the image is limited to one bit per pixel.

Table 3.1 Common raster image file extensions, sources and color depth support

File ext.	Primary source	Color depth 1 grey scale	4	8	16	Support 8 color	16	24	32
ART	First Publisher clip art	×							
BMP	Microsoft Windows RGM encoded	×	×	×		×		×	
BMP	Microsoft Windows RLE encoded	×	×			×		×	
BMP	OS/2 RGM encoded	×	×			×		×	
CAL	Computer-aided Acquisition and Logistics	×	×	×					
CLP	Windows Clipboard	×	×			×		×	
CUT	Dr. Halo					×			
DCX	Multiple PCX Images		×	×		×		×	
GIF	CompuServe	×	×	×		×			
GOE	GOES satellite					×	×	×	×
IMG	GEM Paint	×	×			×			
JPG	Joint Picture Experts Group			×				×	×
MAC	MacPaint	×							
PCD	Kodak Photo CD							×	
PCT	Quickdraw Picture			×		×			
PCX	ZSoft PaintBrush	×							
PCX	ZSoft PaintBrush Version 2	×	×						
PCX	ZSoft PaintBrush Version 3	×	×						
PCX	ZSoft PaintBrush Version 5	×	×			×		×	
PDS	NASA Planetary Images	×		×					
PGM	UNI		×						
PIC	PC Paint	×	×			×			
PIC	IBM Picture Maker					×			
PICT	Macintosh image display	×	×	×	×	×	×	×	×
PNG	Portable Network Graphics	×	×	×	×	×		×	×
RAS	Sun Microsystems Raster files	×		×		×		×	×
RIX	ColoRX Paint					×			
RLE	Microsoft Windows		×			×			
SGI	Silicon Graphics Image files	×		×			×		
TGA	True Vision (Targa)			×		×		×	×
TIF	Aldus Corp Huffman Compressed	×							
TIF	Aldus Corp No Compression	×	×	×		×		×	
TIF	Aldus Corp Packed bits	×	×			×		×	
TIF	Aldus Corp LZW compression	×	×	×		×		×	
TIF	Aldus Corp G3 compression	×							
TIF	Aldus Corp G4 compression	×							
WMF	Windows Metafile	×	×	×		×		×	
WPG	Word Perfect Version 5.0	×	×			×			
WPG	Word Perfect Version 5.1	×		×		×			
WPG	Word Perfect Version 6.X	×	×			×		×	

Table 3.2 Common image file formats

Extension	Description
ART	Black and white clip art images used by Software Publishing Corporation's First Publisher.
BMP	The Microsoft Clipboard and file format which stores images in a bit map format and optionally supports RLE compression.
CUT	Produced by the Dr. Halo paint program and used by other DOS-based paint programs, the CUT file stores pixel data whereas a PAL file stores colour information.
GIF	CompuServe's Graphical Interchange Format stores images using a 12-bit Lempel–Ziv–Welch (LZW) lossless compression technique.
JPG	The Joint Picture Experts Group (JPEG) standardized the storage of images based on the ability of a user to specify the removal of details via a lossy compression method.
PCX	The ZSoft PaintBrush image format uses a run length limited (RLL) lossless compression method.
PICT	The default file format for any image displayed by a Macintosh. PICT supports 1, 2, 4, 8, 16, 24 and 32-bit color depths.
PNG	The Portable Network Graphics (PNG) format was developed as a replacement to GIF due to Unisys requesting a royalty for software supporting the GIF format.
TIF	The Tag Image File Format (TIFF) represents a specification for the storage of images that was jointly developed by Aldus Corporation and Microsoft. Although the copyright is held by Aldus, the specification is in the public domain. TIFF supports five compression methods: one lossy and four lossless.
WPG	The Word Perfect Graphics (WPG) format is primarily used for importing different types of graphics into Word Perfect documents.

BMP

The Bit Map Format (BMP) actually represents a series of graphic file formats supported by Windows and OS/2 for nature graphics. The original format was developed as a general-purpose graphic image format designed to store variable size images with color depths ranging from one to 24 bits. Due to the common parent heritage of Windows and OS/2, both operating systems use this file format, although there are variants both between and within the file format for each operating system.

The Microsoft Windows bit map format includes an option which enables the support of a Run Length Encoding (RLE) data compression scheme. Under RLE repeated runs of pixels are compressed, which can result in a smaller BMP file. Although the use of RLE compression makes you expect that the resulting compressed image will be smaller than a non-compressed BMP image, this is not always the case. A 'noisy' image which has many translations of color depth, object changes and other irregularities may well result in an expansion of storage when compressed using BMP's RLE compression option. This expansion results from the use of special characters required to denote the occurrence of compression and the number of pixels compressed adding to the size of the file when runs of the same pixel color depth are relatively short. Even when a substantial portion of the background of an image has the same composition,

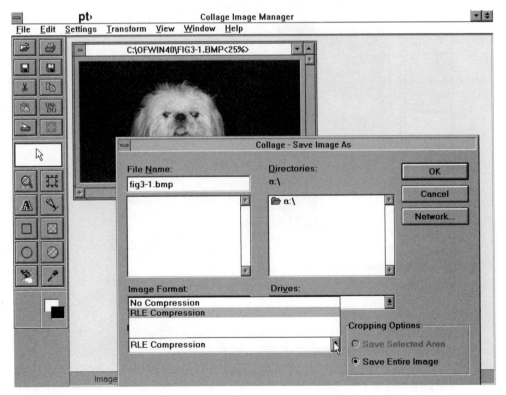

Figure 3.1   Using the Collage Image Manager to create a BMP file using RLE compression.

the use of RLE compression can result in an expansion instead of a contraction of a file. As the proof of the pudding is in the eating, let us use an image manipulation program to convert the previously scanned image of this author's dog into two different BMP files, one file being created using the BMP file format without compression, and the second version of the image will be stored using the BMP file format using RLE compression.

Figure 3.1 illustrates the use of the Collage Image Manager program to convert a file previously stored in one image format to another. In this example, a picture of the author's dog that was previously converted to the BMP format without compression, is now being converted into the BMP file format using RLE compression. Once the conversion was complete, each image was displayed and the Image Information entry in the program's Window menu was selected to display information about the resulting image. This action resulted in the display of a window

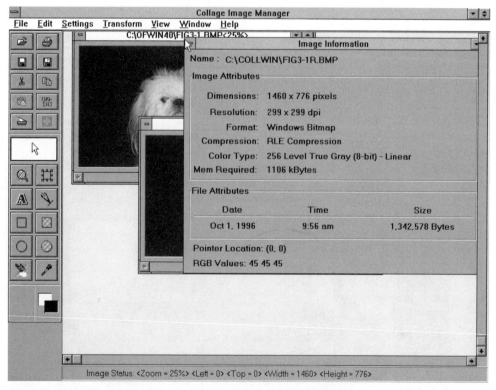

Figure 3.2 Using the Collage Image Manager to display information about a BMP file stored using RLE compression. The use of RLE compression expanded the file which can be noted by comparing its file size to the non-compressed BMP file size shown in Figure 3.3.

for each image labeled Image Information which indicates the attributes associated with each image including the file size, file format and use of compression.

Figure 3.2 illustrates the Image Information window display for the BMP image stored using RLE compression. Note that the file size is 1 342 578 bytes. Figure 3.3 illustrates the Image Information window display for the same BMP image, however, no compression was used when the file was saved. Without the use of RLE compression, the storage required for the file was reduced to 1 134 038 bytes.

It should be noted that the image manipulated into two BMP formats had its greatest variations towards the centre of the image. This eliminated the potential for extremely long runs of the same pixel and reduced the potential effectiveness of RLE compression. However, it should also be noted that for other types of image RLE compression can provide an appreciable amount of data reduction. Because RLE compression is fully reversible, there is no loss of image quality when the compressed data is

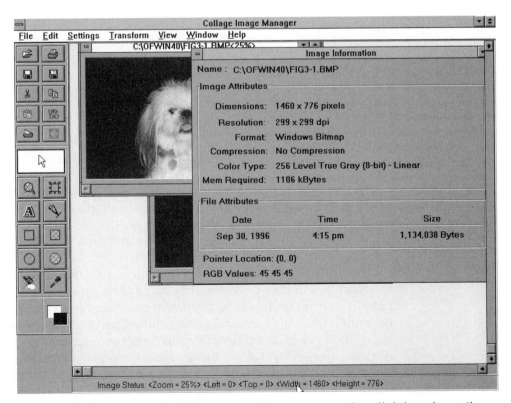

Figure 3.3 The use of a BMP file format without compression slightly reduces the storage required for the image

decompressed and viewed or printed. Thus, RLE compression represents a lossless compression method because all pixels in the original image are restored by decompression.

CUT

The CUT file format represents images created by the Dr. HALO Paint program, and actually consists of two files with the extensions .CUT and .PAL. The .CUT file contains the pixel data associated with the image, whereas the .PAL file contains colour palette information associated with the pixel data. There are two versions of Dr. Halo, one using two bits for color depth and the other using eight bits. The more common version is the 8-bit color depth. Dr. Halo CUT files are used by several IBM PC and compatible based paint programs.

GIF

The CompuServe Graphical Interchange Format (GIF) was one of the earliest image storage and viewing formats, with the first GIF standard, referred to as GIF87a, developed in 1987. A second and downward compatible standard known as GIF89a was developed during 1989. Both formats are widely used by electronic bulletin boards and on Internet Web pages, as they were among the first to incorporate data compression, using a 12-bit Lempel Ziv Welch technique. That compression technique typically provides a 2:1 to 3:1 reduction in the amount of data storage required to store an image, in comparison to storing it in its original bit mapped format.

The key differences between GIF87a and GIF89a are in the areas of multiple image support, interlaced display capability and inclusion of text. Under GIF89a comments can be associated with an image and multiple images can be stored on a file, providing a mechanism to perform a limited but viable animation by storing rotating or displaced sequences of images in a file. Concerning interlacing, GIF89a provides this capability which allows alternate lines from an image to be displayed. This allows an image to be 'built' onto a display rather than slowly 'painted' on the screen. Here the term 'painted' usually refers to the continuous display from top to bottom of all lines of an image in their sequence. As this can take a considerable amount of time when images are transmitted on relatively slow communications facilities, the term can be considered to be derived from the manner in which some people paint a wall from top to bottom.

During 1994 and extending into 1995, a significant degree of controversy occurred concerning the use of GIF. This controversy resulted from a patent dispute between Unisys that holds the LZW compression patent used in the GIF format and vendors developing GIF viewers, image manipulation programs that use GIF, and similar applications. In effect, Unisys considers all software created or modified on or before January 1 1995 that supports GIF and/or the TIFF LZW compression methods to inadvertently infringe on its patent, and will not require developers of such software to obtain a license. However, Unisys expects developers of such software to obtain a GIF-LZW license agreement from that vendor if after December 31 1994 new or modified software is released that supports the GIF file format. Due to this licensing policy, several possible replacements for the GIF format were examined by different imaging users and developers, and a group of persons announced the release of the Portable Network Graphics (PNG but pronounced 'ping') image file format in May 1995. Although many image-based applications now support the PNG image file format, the controversy over the potential for Unisys to collect royalties on the use of LZW has diminished, and the GIF format remains the most popularly used image file format.

To provide a frame of reference concerning the capability of LZW compression, note that the original photograph of the author's dog stored as a TIF file without compression required 1 134 994 bytes. When that image was converted to a GIF file using LZW compression, the size of the file was reduced to 742 763 bytes, a reduction of approximately 35%.

Applications that support BMP and a few other image file formats enable users to enable or disable compression. In comparison, because both versions of GIF automatically support LZW, compression applications do not provide a choice but simply use LZW compression for file storage when the GIF format is selected. Although GIF provides a reasonable level of compression, as we note later in this section, additional data reductions are possible with other image file formats. In spite of this, GIF remains the most popular image file format, based on a random survey of electronic bulletin boards and Web sites performed by this author in early 1997. Part of the reason for this can be attributed to the large number of GIF image manipulation programs that are marketed, as well as the use of GIF89a to provide simple animation, a topic we will discuss in detail later in this book. Similarly to RLE, LZW represents a fully reversible compression method. That is, the decompression algorithm, when applied to a previously compressed file, results in the recreation of the original file on a bit by bit basis without any loss of data. Due to this, LZW also represents a lossless compression method.

JPEG

The Joint Picture Experts Group (JPEG) standardized a method of image storing and viewing based on a series of compression methods. Although the technique is referred to as JPEG, the file extension resulting from an image stored using the JPEG technique is JPG, a carryover from the DOS and non-Windows 95 and pre-Windows NT limitations of a three-character file extension.

JPEG image compression is based on the transformation of 8 by 8 pixel blocks of a true color image into luminance and chrominance levels. The luminance represents the brightness, and can range from zero (pure black) to one for full luminance (pure white). In comparison, chrominance represents the difference between a color and the reference level. Each block is processed by a two-dimensional discrete cosine transformation to obtain 64 coefficients representing the pixels in the block. Those coefficients are quantisized by predefined tables for luminance and chrominance components, after which information about the block is then packed into lower-frequency coefficients. This action results in many coefficients being represented by 0s and 1s, which facilitates the compression of data representing the image.

Many imaging programs that support JPEG enable a user storing an image to adjust the quantization tables by defining a quality value. Unfortunately, there is no standardization concerning the use of a quality value, and a setting used with one program may produce different results when the same setting is used with a different program. Although most programs use a scale from 0 to 100, some programs use predefined quality values that result in the user interface restricted to high, medium, and low style choices, and other programs use a scale of 1 to 4 or 1 to 5. As the quality setting in a program affects the manner in which blocks of pixels are compared to one another, differences in the implementation of quality settings and quality scales do not prevent different implementations from exchanging JPEG files. Thus, differences in the use of quality values do not affect the ability to use JPEG files created by one acquisition method in other image manipulation programs, or between image manipulation programs.

In most image manipulation programs that support a JPEG quality value scale of 0 to 100, a setting of 75 is often the best choice. This setting usually results in blocks of pixels being considered equivalent with a minimal amount of difference between blocks. Thus, differences between the original image and the image produced by a quality value below 75 begin to become apparent as you use values below 75. In fact, at the default of 75 used by most programs, very little image degradation occurs; however, a significant amount of compression may be obtainable. By returning to this author's favorite pet, we can examine the effect of using JPEG with different quality values.

To obtain the ability to directly adjust the compression quality, the program CJPEG developed by the Independent JPEG group was used. This program is part of a program pair, with the second program DJPEG providing a mechanism to convert images from JPEG to other file formats. Figure 3.4 illustrates the basic program help screen, displayed by entering the program name by itself.

As indicated by the CJPEG help screen shown in Figure 3.4, you can use the −quality switch (abbreviated as −q) followed by an integer between 0 and 100 to specify the quality of the image; however, as we will note later in this section, the quality scale used by image manipulation programs are not standardized, and the setting used by one program can produce different results from the use of the same setting with another program.

A second setting of the CJPEG program that deserves a degree of elaboration is the Huffman setting and the omission of the support of arithmetic compression. The JPEG specification defines

```
C:JPEG>cjpeg
C:\JPEG\CJPEG.EXE: must name one input and one output file
usage: C:\JPEG\CJPEG.EXE [switches] inputfile outputfile
Switches (names may be abbreviated):
    −quality N      Compression quality (0...100; 5–95 is useful range)
    −grayscale      Create monochrome JPEG file
    −optimize       Optimize Huffman table (smaller file, but slow
compression)
    −targa          Input file is Targa format (usually not needed)
Switches for advanced users:
    −restart N      Set restart interval in rows, or in blocks with B
    −smooth N       Smooth dithered input (N=1..100 is strength)
    −maxmemory N    Maximum memory to use (in kbytes)
    −verbose or -debug   Emit debug output
Switches for wizards:
    −qtables file   Use quantization tables given in file
    −sample HxV[,...]   Set JPEG sampling factors

c:\JPEG>
```

Figure 3.4   The Convert to JPEG (CJPEG) Program Help Screen

the use of two 'back end' compression modules for the output of compressed pixel blocks, Huffman cooling or arithmetic coding. The choice of compression method has no effect on image quality, and only serves to reduce the resulting file containing the image. Huffman coding is in the public domain and is usually included in all JPEG implementations. In comparison, arithmetic coding is subject to patents owned by several companies and cannot be incorporated into a program without a license from those companies. Although the use of arithmetic coding can produce a file 5–10% smaller than obtained from the use of Huffman coding, most implementations of JPEG only implement the public domain Huffman coding compression method. As we require a frame of reference the GIF image of the author's dog, which was illustrated in Figure 3.3, will be used. As previously noted, this image required 742763 bytes of storage, with its data storage requirements reduced by the use of LZW lossless compression. A full version of the image is shown in Figure 3.5.

The execution of CJPEG is based on command line entries. As an initial test of the use of CJPEG a quality of 50 was selected, resulting in the following command line entry:

$$CJPEG - q \times 50 \; GIZM.GIF \; GIZMO50.JPG$$

In the preceding command line entry, the target or destination file had the suffix added to denote the quality used for the conversion. The resulting image required 61721 bytes of storage and a hard copy display of the image is shown in Figure 3.6.

Figure 3.5   A picture of the author's dog after conversion to the GIF file format

Figure 3.6   The author's dog displayed after a conversion to JPEG using a quality factor of 50

In comparing Figures 3.5 and 3.6, it is difficult for the naked eye to note any difference between the two; however, with respect to their data storage requirements, there is a significant difference. In this example, the JPEG image requires less than one-twelfth of the storage of the GIF image!

Continuing our examination of the effect of lowering the quality level for subsequent conversions to JPEG, the CJPEG program was re-executed using quality values of 20, 10 and 5. The results of these subsequent conversions to JPEG are illustrated in Figures 3.7–3.9.

Figure 3.7   Displaying the author's dog after a conversion to JPEG using a quality factor of 20

Figure 3.8   You can see numerous 8 by 8 pixel blocks when the JPEG quality factor is reduced to 10

In examining Figure 3.7, if you have an eagle eye you can see that distorted blocks of pixels have begun to appear on old Gizmo's body and in areas in the background, because blocks with more than a few differences in composition are assumed to be equal for compression, and on decompression cannot be used to restore the image to its original form. In examining Figure 3.8, which used a quality value of 10, the pixel blocks are more pronounced, whereas the use of a quality value of 5 shown in Figure 3.9 has distorted the image of old Gizmo to the point where he may no longer talk to his

Figure 3.9   When the JPEG quality factor is reduced to 5, the distortion of the resulting JPEG image is very pronounced

Table 3.3 Comparing GIF to various JPEG quality factors.

GIF	LZW compression	742 763
JPEG	q = 75	112 078
JPEG	q = 50	61 721
JPEG	q = 25	33 277
JPEG	q = 10	23 721
JPEG	q = 5	20 951
JPEG	q = least	871 948
JPEG	q = moderate	71 934
JPEG	q = high	36 339

owner! In all fairness, it should be pointed out that the CJPEG program displayed the statement 'Caution: quantization tables are too course for baseline' when the quality value was set at 20 and below as a well-advised warning. Although you can continue to reduce the size of JPEG files by using smaller quality values, usually somewhere below 50, distortion begins to build up that can significantly affect the quality of an image. Table 3.3 provides a comparison of the results of storing the image of the author's dog in GIF and several types of JPEG files, with the latter resulting from either the direct or indirect variance of the quality values.

In examining the entries in Table 3.3 the least, moderate and highest quality (q) settings represent settings in Collage Image Manager. That program, as well as other image application programs, does not permit a user to directly alter the JPEG quantization tables by defining a quality value. Instead, they internally define several quality values and associate those values with terms like least, moderate and highest or similar descriptors. Then selecting a compression level descriptor results in the program using a predefined quality value.

It should be noted that the use of a low quality value can impact the ability of some programs to work with a resulting converted image. For example, Collage Image Manager would not open a JPG file created using a quality factor at or below 20, resulting in this author using Paint Shop Pro to view and print the resulting files.

Based on the JPEG quality value settings summarized in Table 3.3, and the preceding series of modified photographs of the author's dog, we can develop some general rules of thumb. First, for good quality you should first use a quality value setting of 75, which will provide a significantly reduced file in comparison with the use of a GIF file format. Secondly, as you move up the scale the file size can significantly increase, expanding to a size beyond that required by the use of GIF. This is illustrated by the q = least entry in Table 3.3, which is equivalent to a quality setting of 95 and results in a file size that exceeds the size of a GIF file by approximately 130 kbytes and a JPEG file with a quality value of 75 by approximately a factor of eight! A third rule of thumb is the fact that very low quality values, which produce significantly shrunken files, also produce severely degraded images. Although a very small file resulting from a quality setting at or below 10 might be suitable for an image preview or indexing purpose, you should probably avoid those values for general use.

As indicated by the series of quality value variances, the use of JPEG can provide a substantial data reduction while maintaining good image quality when a quality value at or above 50 is used. Owing to this, JPEG has become the second most popular image file format used on electronic bulletin boards and Web pages. A recent addition of the capability to display interlaced JPEG images can be expected to further increase the use of this image format on World Wide Web pages. Owing to JPEG's excellent compression ratios when applied to photographs, we can expect its use on Web pages to increase in tandem with the growth in the popularity of digital cameras that are being used to 'spice up' the contents of those pages through the addition of pictures of persons, places and things.

PCX

The PCX raster image format was developed for the Zsoft PC PaintBrush image editing program, and it represents one of the first graphics formats used with the IBM PC and compatible computers. The PCX format is supported by many DOS and Windows-based programs and it supports color depths ranging from black and white to 24-bit true color.

PCX used a run length compression method which results in a repetitive sequence of bits being replaced by a repeat byte count and a data byte.

Although the PCX image format incorporates data compression as a default, the method of compression used is similar to the BMP's RLE method in that it is not terribly efficient. To provide readers with a frame of reference, old Gizmo's picture was converted to a PCX format. In doing so the resulting image required 1 207 948 bytes of storage, which exceeded the Windows BMP image format without compression (1 134 038), but which was less than the BMP image format when RLE compression (1 342 578) was employed.

The run length compression method used by PCX is relatively fast, enabling coding and decoding to occur essentially 'on the fly' with disk access more of a delay than processing time. However, when comparing this lossless compression method to GIF's LZW method and several types of lossless compression used in the PNG and TIFF image formats described later in this chapter, it becomes obvious that the other image formats provide better storage and transmission results.

PNG

The license problems associated with Unisys patent on the LZW data compression technique, and its attempt to extract royalties from software developers using GIF, resulted in the motivation for a substitute for the GIF image format. The result was a similar but legally unencumbered standard for lossless bitmapped image files known as the Portable Network Graphics (PNG) specification.

PNG was initially designed as a substitute for the GIF89a format, using Lempel–Ziv–Huffman (LZHUF) data compression in place of LZW. As LZHUF was in the public domain, this substitution enabled image application software developers to use PNG without having to consider the payment of royalties for the incorporation of the support of an image file format. Although PNG includes most key GIF89a features, such as the support of up to 256 colors, interlacing on the progressive display of an image and the use of lossless compression, it added several features beyond GIF. Those new features include gray scale support up to 16 bits per pixel and truecolor support up to 48 bits per pixel, as well as a mechanism which provides a faster initial presentation of the image in a progressive display mode.

The LZHUF compression method used in PNG provides a significant data reduction beyond that obtainable through the use of GIF. To illustrate this fact the author again turned to the picture of his dog, using an image conversion program to convert the GIF file format to a PNG file format. As previously noted, the GIF file format required 742 763 bytes of storage. When the GIF image was converted into a PNG file, the resulting file was reduced to G74588 bytes. As both GIF and PNG use lossless compression, the images stored in each format are visually identical.

Although PNG provides a superior data reduction capability to GIF in its present release it does not support multiple images in a file. This omission is extremely important for many Web page developers, as GIF89a's support enables a degree of animation to be added to Web pages. Owing to PNG's lack of support of multiple images in a file and its relatively newness, this image file format may require a period of time and the incorporation of direct support into Web browsers before it obtains a significant degree of popularity.

TIF

The first Tag Image File Format (TIFF) specification was jointly developed by Aldus, Microsoft and several scanner manufacturers during 1986 as a mechanism to standardize a file format for images used in desktop publishing. Since then, a number of specification revisions have occurred which have expanded TIFF support to digital video images, and have increased the number of compression methods that can be used to reduce image storage requirements.

TIFF supports five compression methods, four lossless and one lossy. Lossless compression methods supported include two types of ITU-T (formerly known as the CCITT), Group 3, and one method of Group 4 compression primarily associated with the use of fax, LZW and Huffman compression. A lossy JPEG compression method was added in TIFF specification revision 6.0. A TIFF compatible file generator program must support at least one compression method, whereas a TIFF compatible reader should be capable of supporting all compression methods. Similarly to JPEG, TIFF files are commonly stored using a three-character extension, which explains why the second F is not used.

The Group 3 compression method represents the use of a one-dimensional modified Huffman run-length-encoding technique that was developed to facilitate the transmission of black and white faxes. Under this technique, runs of black or white pels (picture elements), which is a term used to represent pixels on a fax, are compared to predefined tables of runs. When there is a match, the predefined Huffman code is substituted. In the Huffman coding technique the most common values are assigned the shortest codes, and the least frequently occurring values are assigned the longest codes. The Huffman codes assigned to pel runs are based on an examination of common documents and are static. As Group 3 fax is based on a one-bit color depth, you can reasonably expect to lose clarity when converting a photograph to a Group 3 fax TIF file format. This is illustrated in Figure 3.10 which shows the resulting conversion of the image of the author's dog to a Group 3 TIF format. Although the resulting data storage was reduced to 60 146 bytes, its loss of detail makes the Group 3 format more suitable for drawings and text than photographs.

The use of the TIFF Group 4 compression option produces a result very similar in both file size and image clarity to the use of Group 3 compression. That is, the use of Group 4 also results in a decrease in color depth, and the resulting file size obtained by the conversion of the photograph of the author's dog was reduced to 60144 bytes.

The use of LZW compression produces an image similar in both appearance and size to the use of the GIF file format. The photograph of the author's dog is indistinguishable from that of the GIF image shown in Figure 3.5, and the file size was reduced to 752 598, approximately 10 000 bytes larger than the GIF file. Two other TIFF compression options were also employed to note the differences between using each TIF option: no compression and JPEG. When no compression was used the file size was 1 134 994, and the use of a quality value of 50 resulted in a JPEG file size of 61 848 bytes which very slightly exceeded the size of the file produced by the CJPEG program, with the difference in storage most likely attributable to the difference in headers used

Figure 3.10 The Group 3 TIFF format results in a decrease in color depth that considerably alters the details of photographs

Table 3.4 Comparing TIFF compression options

Option	Storage
No compression	1 134 994
LZW	752 598
JPEG	61 848
G3	60 146
G4	60 144

by each file format. For comparison purposes, Table 3.4 summarizes the various compression options of TIFF applied to the photograph of the author's dog.

WPG

The Word Perfect Graphics (WPG) format is primarily used to import different types of Graphics into a Word Perfect document. WPG supports both bit maps (raster) and vector formats; hence, it can be considered as a metafile format.

WPG supports a variety of color depths, such as black and white, 16 level True Gray (4 bit), 256 level True Gray (8 bit), 16 color (4 bit), and 256 color (8 bit); however, there are differences between versions of Word Perfect with respect to the color depth of WPG images supported. The primary difference involves the use of 256 color files. If you are using a version of Word Perfect that does not support 256 color WPG images you can consider using an image manipulation application program to either reduce the color depth to 4 bits (16 colors) or to dither the image to monochrome.

Figure 3.11 illustrates the placement of a WPG image into a Word Perfect document. The actual conversion of the image into a WPG image format resulted in a file size of 1 185 976 bytes, which is slightly larger than a non-compressed TIF image. When that file is incorporated into the Word Perfect document shown in Figure 3.11, the difference between a text-based file and one with one or more graphics is enormous. Instead of having a word-processing document that requires perhaps a few thousand bytes of storage to describe the author's dog, the use of the picture of the pet explodes the file to well over 1 Mbytes of storage. Thus, the use of images within such common applications as wordprocessing documents is increasing the storage and transmission time of such documents by a factor of 1000 or more, again illustrating the need to consider carefully the manner in which images are supported on a network.

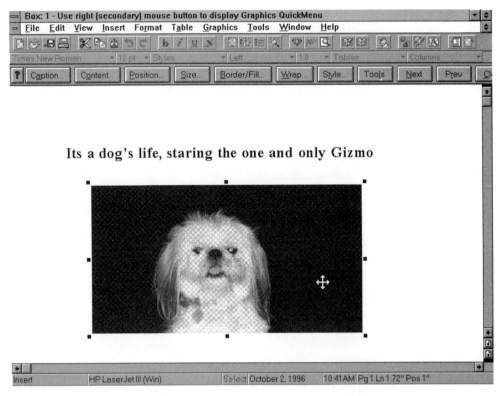

Figure 3.11   Incorporating a WPG image into a Word Perfect document

Format comparison

In concluding this section covering raster images, we will use the series of previously performed image format conversions to summarize the storage requirements associated with the use of different file formats and compression options supported by certain formats. This summary, which is included in Table 3.5, deserves some words of caution. This is because the conversions were based on the use of one image, and differences between the results summarized in Table 3.5 will vary when another image is used for comparison purposes. However, it should also be noted that although some images may enable BMP's RLE compression to be more effective, for most photographs the results will be similar. That is, JPEG will normally provide the smallest file size with an acceptable level of clarity with a quality value of 50. If lossless compression is required, PNG will normally provide a greater data reduction than GIF; however, prior to selecting an image file format it is important to consider the application to be used. For example, to use PNG with currently available Web

Table 3.5 Image file format storage comparison

BMP	no compression	1 134 038
BMP	no compression	1 342 578
GIF	LZW compression	742 763
JPEG	q = 75	112 078
JPEG	q = 50	61 721
JPEG	q = 25	33 277
PCX	run length	1 207 948
PNG	LZHUF compression	674 588
Targa	standard compression	1 132 978
TIF	no compression	1 134 994
TIF	LZW	752 598
TIF	JPEG	61 848
TIF	G3	60 146
TIF	G4	60 144
WPG	no compression	1 185 976

browsers requires a plug-in module PNG viewer. In comparison, most browsers directly support GIF and JPEG images. Thus, you may have to weigh the availability and cost of programs or plug-in modules to display and print different image formats.

3.2 VECTOR IMAGE FORMAT BASICS

As noted at the beginning of this chapter, there are two primary types of images: raster and vector. A vector image is based on algorithms that draw lines, arcs, polygons and other objects. Thus, a vector file commonly represents values used by algorithms to produce an image, such as the starting and ending coordinates for a line, its color and width. The most popular applications that use vector images are graphs, charts and simple-to-complex diagrams.

Currently there are a large number of vector image file formats, with each format primarily supported by a specific application program. In an attempt to add a degree of standardization to the use of vector images, a new vector graphics format was developed by the vendor SoftSource and the National Center for Supercomputer Applications (NCSA). Known as the simple vector format (SVF), this vector graphics format was developed to provide a 2-D vector format for use on the World Wide Web.

The simple vector format

SVF supports points, lines, polylines, rectangles, circles, arcs, cubic bezier curves and text. The format also enables the inclusion of hyperlinks in each object, allowing the graphics format to support the operation of browser point and click movements. In fact, an SVF plug-in for the Netscape browser has been developed which enables SVF images to be embedded in HyperText Markup Language (HTML) documents placed on Web servers and viewed through the use of Netscape.

Other popular vector formats

If you work with a drawing program as opposed to a paint program, the files that you save are stored as vector images. Five popular vector image formats are listed in Table 3.6 including their file extension. In examining the entries in Table 3.6, note that WPG more accurately represents a metafile, as it supports both raster and vector image formats.

Compression considerations

As vector image files in effect represent compressed images, most programs that generate such files do not include a compression option. Although you can use an independent compression-performing program to compress a vector image file, the results are usually not worth the effort unless you can either automate the process or have a large number of such files to transmit on a recurring basis. Both these topics will be examined later in this book.

Table 3.6 Popular vector image formats

Extension	Description
CDR	CorelDRAW!
DRW	Micrografx Draw
DXF	Autodesk
HGL	Hewlett-Packard Graphics Language
WPG	Word Perfect

IMAGE ACQUISITION TECHNIQUES

Until now we have essentially focused our attention on the technical details of a variety of image file formats and the effect on storage of converting from one file format to another. As noted in the introductory portion of the preceding chapter, our focus on image acquisition techniques was purposely delayed to obtain the ability to discuss such techniques with respect to different file formats. Thus, having obtained a foundation concerning the technical details associated with images and their file storage requirements resulting from the use of different file formats, we will now turn our attention to image acquisition techniques.

In this chapter we will focus on different methods by which images are input into a computer for manipulation. In doing so, we will discuss options provided by several techniques, as well as the data storage requirements and, when applicable, the transmission time associated with each technique.

In general we can categorize image acquisition techniques by their relationship to a computer. That is, those techniques either require the direct presence of a computer via the cabling of an input device to the processor, or they depend on communications. Thus, in our examination of image acquisition techniques, we will first focus our attention on a group of techniques which have a degree of dependence on the use of an input device directly connected to the processor. This will be followed by an examination of techniques that use a transmission facility.

4.1 DIRECT ACQUISITION METHODS

There are a variety of image acquisition methods that directly place files storing images into a computer. Such techniques range

Table 4.1   Direct image acquisition
techniques

Diskette transfer
Application Program creation
Digital camera
Screen capture program
Screen grabber

in scope from images generated by application programs to images generated by such hardware devices as video image grabbers and scanners. Table 4.1 lists five direct image acquisition techniques.

Diskette transfer

Although the transfer of images into a computer via a diskette may at first glance appear to be so obvious that it does not warrant a discussion; as an old adage notes: 'The obvious is not always the obvious.' Today the majority of diskette drives have the same 1.44 Mbyte capacity as those introduced approximately 12 years ago. Although Iomega's popular Zip drive provides a 100 Mbyte storage capacity, it is not backward-compatible with 1.44 Mbyte drives and forces many computer users to use the older, conventional, lower capacity drives to create diskettes for the distribution of data. This means that for the foreseeable future a majority of image file distributions via diskette will continue to occur via the use of a $3\frac{1}{2}$ inch 1.44 Mbyte media.

Media storage problems

When distributing images using 1.44 Mbyte capacity diskettes, one of the first problems that you will note is the limited capacity of such diskettes to store more than a handful of images. In fact, depending on the image and its storage format, you may not be able to store even one image on a 1.44 Mbyte diskette. However, because that disk media still represents a common denominator for distributing data, techniques have been developed to enable large images to be spanned over multiple disks as well as placing additional images onto a diskette. Such techniques involve the use of compression and archiving programs, such as the popular PKZIP program series from PKWARE and ARJ developed by Robert Jung.

Using compression and archiving

To illustrate the advantages associated with the use of a compression and archiving program, we will use the PKZIP program with several images previously discussed in Chapter 3. Prior to doing so, let us review the format of the use of PKZIP which is shown below:

PKZIP switches Zipfile Source file(s)

In the preceding format, you can include one or more optional switches which govern the operation of the program. For our first example, let us assume we wish to transfer the non-compressed TIF image and the compressed GIF image of the author's dog. The non-compressed TIF file requires 1 134 994 bytes of storage, and the compressed GIF file requires 742 763 bytes of storage. Thus, together they would require 1 877 757 bytes of storage which would exceed the capacity of a 1.44 Mbyte diskette.

Through the use of PKZIP we can significantly compress the non-compressed TIF file. In addition, even though the GIF file format results from the application of LZW compression, there may be some redundancies in the compressed file that can be further reduced. To illustrate the effect obtained from using PKZIP, let us first use the program. Thus, we would enter the following command to compress the two image files:

PKZIP GIZMO.ZIP GIZMO.*

In the preceding command line entry, the use of the asterisk results in any file named GIZMO, regardless of its extension, being compressed and archived onto the file GIZMO.ZIP. We can note the effect of the use of the PKZIP compression and archiving program by using the program with its −v switch to view the contents of the archive we just created. Figure 4.1 illustrates the use of the −v switch to view the contents of the archive.

In examining the contents of the archive displayed in Figure 4.1, note that the column labelled 'ratio' indicates the percentage of data reduction for each file in the archive. Also note that the GIF file that was constructed using LZW compression was further reduced by 1%, which, although not significant by itself, illustrates an important concept. That is, often it is possible to compress previously compressed images to obtain further data reductions that can be significant for both data storage and data transmission. To illustrate this important concept, the picture of the author's dog was converted to a JPEG image using Collage Image Manager and specifying the highest level of compression. The

C:\JPEG>c:\pkzip\pkzip −v gizmo.zip

PKZIP (R)   FAST!   Create/Update Utility   Version 2.04 g   02-01-93
Copr. 1989-1993 PKWARE Inc. All Rights Reserved. Shareware Version
PKZIP Reg. U.S. Pat. And Tm. Off.   Patent No. 5,051,745

* 80486 CPU detected.
* EMS version 4.00 detected.
* XMS version 2.00 detected.
* Novell Netware version 3.11 detected.
* DPMI version 0.90 detected.

Searching ZIP: GIZMO.ZIP

Length	Method	Size	Ratio	Date	Time	CRC-32	Attr	Name
1134994	DeflatN	555877	42%	10-01-96	11:11	d143c9c7	- - w -	GIZMO.TIF
742763	DeflatN	738669	1%	10-04-96	08:31	5c7a6ce0	- - w -	GIZMO.GIF
1877757		1405546	26%					2

C:JPEG>

Figure 4.1   Using the PKZIP −v option to view the contents of the archive stored on the file

resulting JPEG file required 36 561 bytes of storage which is significantly less than the TIF and GIF formats. However, the fact that JPEG performs compression on 8 by 8 pixel blocks and makes such blocks equivalent when they differ by a few pixels as you change the quality value also makes the resulting file suitable for recompressing using a lossless compression technique, such as the one built into PKZIP. This is illustrated by again using PKZIP to add the JPG file to the archive and once more using the program's −v option to view the contents of the modified archive. Figure 4.2 illustrates the contents of the revised archive. In examining Figure 4.2, note that PKZIP compressed the JPEG image stored as GIZMO.JPG by 22%.

As the file storage requirements of individual or multiple files exceed the capacity of a single diskette, you can consider forming an archive that spans multiple diskettes. PKZIP and many similar archiving and compression programs include a disk spanning feature. If you are using PKZIP, that feature is invoked through the use of the program's −& switch. Thus, the following command would be entered to compress and archive a series of images on a directory on your hard drive onto a series of diskettes:

PKZIP −& IMAGE.ZIP *.*

```
C:\JPEG>c:\pkzip\pkzip −v gizmo.zip

PKZIP (R)     FAST!     Create/Update Utility     Version 2.04g     02-01-93
Copr. 1989-1993 PKWARE Inc. All Rights Reserved. Shareware Version
PKZIP Reg. U.S. Pat. And Tm. Off.     Patent No. 5,051,745

* 80486 CPU detected.
* EMS version 4.00 detected.
* XMS version 2.00 detected.
* Novell Netware version 3.11 detected.
* DPMI version 0.90 detected.

Searching ZIP: GIZMO.ZIP

Length    Method   Size      Ratio   Date       Time    CRC-32    Attr    Name
------    ------   ----      -----   ----       ----    -------   ----    ----
1134994   DeflatN  555877    42%     10-01-96   11:11   d143c9c7  - - w -  GIZMO.TIF
 742763   DeflatN  738669     1%     10-04-96   08:31   5c7a6ce0  - - w -  GIZMO.GIF
  36561   DeflatN   28521    22%     10-04-96   09:05   32509eb0  - - w -  GIZMO.JPG
------             ------    ---                                          ------
1914318            1434067   26%                                              3

C:JPEG>
```

Figure 4.2   Viewing the contents of the updated archive

When creating an archive that spans multiple diskettes, PKZIP can be used with another switch to format diskettes as it prompts you to insert different diskettes. This is a most convenient feature, that is truly appreciated in the event that you run out of formatted disks without completing the transfer of an archive and need to restart the operation once you terminate the operation and format the additional diskettes required for the spanning operation.

As noted in this examination of diskettes as a method for acquiring images, the size of an image should not by itself eliminate this method from consideration. If you have or acquire a compression and archiving program with a diskette spanning capability, you can place one large image on multiple diskettes. In addition, through the use of the compression capability of the program, you may be able to significantly reduce the size of one or more files that enables them to be stored on one or a few disks.

Application program creation

Today you can create a variety of images through the use of different types of application program, such as paint and drawing program, as well as electronic spreadsheets and charting programs that can generate a variety of graphs. When considering the use of application programs to create images, it is important to consider the file format produced by the program in conjunction with the intended use of the image. If the file format is not compatible with the format required and you cannot locate a conversion program that supports both formats, do not despair. Two additional options that deserve consideration are the use of a screen capture program and the use of a scanner.

Conversion options

If you can display the image, you can use a screen capture program to store the image. Most screen capture programs support a number of commonly used image formats, enabling you to capture a chart, graph or another object that may be stored in a proprietary format that would otherwise preclude its use. For example, consider Microsoft's PowerPoint program that can be used to generate overhead transparencies and slides. Although the program stores data to include imported images in a proprietary format, you could use a screen capture program to store a desired slide in a common format, such as GIF or JPEG, that could be used on a Web page.

If you can print an image, you can consider using a scanner to input the image into your computer. Similarly to most screen capture programs, most software provided with scanners supports a number of common file formats. Later in this section we will examine the use of screen capture programs and scanners.

Digital cameras

The digital camera holds the promise to become as common a computer peripheral as the ubiquitous printer. Based on the rapid increase in the use of digitized images incorporated into wordprocessing documents, World Wide Web server home page documents, and databases with pictures of buildings and employees, it is expected that within a few years many persons will replace conventional cameras by digital cameras because their output can be directly used by a computer. By bypassing the requirements to first develop photographs and then obtain and use a scanner to digitize photographs so they are suitable for use with computers, the use of a digital camera can result in a saving of time and money, and enhanced user productivity.

Operation

A digital camera is based on the use of Charged Coupled Devices (CCDs), similar to the CCDs used in many types of scanners. A CCD converts light energy into an electric current and is fabricated as a matrix of R rows and C columns to provide an array of $R \times C$ elements.

Each element in a CCD array provides an analog signal in the form of a voltage which varies in proportion to the spectrum of white light seen by the element. Through the use of an electronic analog to digital converter, the voltage output of each CCD element is converted into a group of bits which represent the digitized value of the element. The digitized values of all the elements from the CCD array are stored in the camera's RAM, permitting a user to download images from the camera directly into a computer, bypassing previously required photographic development and scanning processes.

The response of each CCD element to light differs from film which uses grains of silver oxide. As film has a much higher density of grains of silver oxide than a CCD, film provides a higher resolution than obtainable through the use of current technology incorporated into digital cameras. In addition, the random grain structure of silver oxide used to produce film provides a 'warmer' feeling image as CCD devices tend to provide a harder, grittier quality image. Both scanned images and the output of digital cameras provide a more static image which to the eye may appear a bit coarser than an image captured on film. However, in many instances the output of a digital camera, although not providing photographic quality, may be more than sufficient for the incorporation of images into visual databases, brochures, newsletters and other types of documents.

Types of digital camera

There are several types of CCD-based digital camera, ranging from very near 35 mm quality cameras that can cost in excess of $30 000, to poorer resolution but still relatively good quality cameras that you can obtain for under $500. Even if you have available funds to purchase a top of the line digital camera it may not be practical to do so. This is because the ability to use the higher resolution of more expensive digital cameras is dependent on the capability of your hardware platform. For example, higher-resolution images require additional data storage which can tax the capacity of your hard drive and limit the number of images that you can store. In addition, if your monitor and laser printer do not have an acceptable level of resolution, you will not be able to view nor print the image at its recorded level of resolution. Thus, it is important to consider your monitor and laser printer levels of resolution, as well as the number of images that you will store on your hard disk, the storage requirements of each image, and your available hard disk storage capacity prior to selecting a digital camera. Once this has been accomplished, you can focus your attention on the different types of digital cameras currently being marketed. In doing so you match the characteristics of different types of camera against the capabilities of the hardware platform that you anticipate using with the camera.

Currently there are three types of digital camera, with each type primarily based on the method by which the CCD built into the camera operates. A slow scan digital camera includes a built-in linear CCD which makes a series of passes on a large array of CCD elements to capture an image. A small CCD, which moves to different quadrants of an array to capture an image, is referred to as a stepping chip. A third type of CCD uses a one-dimensional CCD array. Known as a linear array CCD, the use of a one-dimensional CCD array enables a larger area of an image to be captured

A slow scan digital camera is based on the use of a built-in large CCD which operates in a linear manner. The use of a large CCD containing many elements results in a relatively long exposure time to capture an image. Most digital cameras that use a built-in linear CCD are designed to use a three-pass exposure process similar to a color scanner. Here each exposure is for one of the primary colors, resulting in a true RGB color image being captured.

The linear CCD based digital camera is the most expensive type of camera marketed; however, this camera produces the highest resolution currently obtainable. One representative camera in this category is manufactured by Leaf Systems as an adapter that replaces the back of a Hasselbdad or Mamiya camera and produces an image stored as a 42-bit proprietary file consisting of a 14-bit color depth for each of the primary colors. When this file is imported into a PC, it provides an image with a 2048 by 2048 resolution with a true color 24-bit color, requiring approximately 12 Mbytes of storage. When images are captured using the camera's 8-bit gray scale, an image with a resolution of 2048 by 2048 requires approximately 4 Mbytes of storage.

Although the linear CCD-based digital camera provides an image whose quality is near to if not visually equivalent to film, its cost may be prohibitive for most potential users. Currently, the cost of a linear CCD-based digital camera is approximately $35 000 to $40 000. In addition, unless you intend to use the image obtained from a linear CCD-based camera in a high-quality magazine or book, its resolution and storage requirements are probably beyond that needed nor wanted for most other applications. For example, would you want to store a 4 or 12 Mbyte image on a popular Web server? If you did, each time the image were to be retrieved it would consume such a significant amount of network bandwidth that its downloading would adversely effect other users attempting to access the server.

A stepping chip digital camera is based on the use of an inexpensive low-resolution CCD to generate high-resolution images. To accomplish this the stepping chip camera takes several low-resolution images. The camera then splits the images into quadrants and then processes each quadrant, once for each primary color, to create a composite high resolution image.

A stepping chip based digital camera is limited to a resolution of 2048 by 2048 and can require from 15 seconds to one minute to capture and process an image. Although the camera does not provide as good a quality image as a slow scan camera, its price can be between 20 and 40% less, commonly ranging between $20 000 and $30 000. As most PCs and laser printers cannot take advantage of the high resolution of a stepping chip camera, this type of camera may be more than sufficient for most office type applications, unless you anticipate capturing images that will be used with typeset documents and similar commercial operations that require the higher resolution obtainable from the use of a slow scan camera.

Two examples of stepping based digital cameras are the Kodak DCS 200 MI and DCS 420. The DCS 200 camera is actually a base unit containing a monochrome CCD that is fitted to a Nikon 8008 camera's film plane. By making three exposures through a red, blue and green filter wheel, a true RGB color is obtained. This digital camera provides a 24-bit RGB image at a resolution of 1524×1012, but uses an internal hard drive for image storage which limits the transfer of image to cabling between the camera and computer. The more modern DSC 420 is an electronic back fitted to the body of a Nikon N90 camera. This digital camera has the same resolution of the earlier developed DCS 200, but includes several features that facilitate its use and were incorporated into lower cost and lower resolution products. The DCS 420 stores images on a Type III PC Card which enables captured images to be easily moved into a computer or notebook that includes a Type III PC Card slot. The camera also includes an SCSI interface that supports the high-speed downloading of stored images into a computer when you cannot take advantage of its Type III PC card. If you are using the camera in the office, you can also capture images directly to your hard drive when the camera is cabled to your computer.

A linear array based digital camera sacrifices some areas of performance to provide economical cost which can range from approximately $10 000 to under $500, based on the level of image resolution produced. As the image resolution is a function of the CCD array size, a lower resolution linear array based digital camera results in a reduction in the number of CCD elements in the CCD array, lowering the cost of the most expensive part of the camera.

A linear array based camera operates similarly to the CCD incorporated into a flatbed scanner. That is, a one-dimensional array is used to repeatedly scan different areas of a desired image. Although the reduced CCD array results in a significant reduction in the cost of the camera based on this technology, it introduces several performance tradeoffs. First, the scanning process is relatively slow in comparison to conventional CCD matrix based digital cameras. This means that the camera and/or object being photographed cannot move for a short period of time. If either the camera or object moves during the linear scan process, the image will appear blurry. Thus, this type of camera is not designed for use at sporting events nor for use where precious moments involve movement.

There are two methods used with linear scanning digital cameras. The first method involves the use of a three-pass technique with internal RGB filters. The second method involves the use of a single pass with the use of three linear CCDs, each one functioning as a separate RGB filter. Although a three-pass linear scanning CCD based camera is slower than a single-pass camera that uses three linear CCDs, it is considerably less expensive. In fact, most low-cost digital cameras that have a retail price under $1000 are based on this type of CCD, and during 1997 several cameras reached the market with a retail cost under $500.

Camera features

In addition to the type of CCD used by a digital camera, there are several other features that you should consider during the camera evaluation process. Table 4.2 lists 11 features that you should consider when evaluating digital cameras, including the previously discussed type of CCD used in the camera.

Table 4.2   Digital camera evaluation features

Feature	Requirement	Vendor A	Vendor B
type of CCD used	_____	_____	_____
focus area	_____	_____	_____
computer interface	_____	_____	_____
image resolution	_____	_____	_____
color depth	_____	_____	_____
camera RAM	_____	_____	_____
camera image storage capacity	_____	_____	_____
image compression	_____	_____	_____
bundled software	_____	_____	_____
cost	_____	_____	_____
warranty	_____	_____	_____

In examining Table 4.2, note that it is structured as an evaluation worksheet on which you can list your specific camera requirements for each feature. Then you can compare and contrast the features of two vendor products against your requirements. If you wish to evaluate more than two vendor products you can extend Table 4.2 to the right and add an appropriate number of columns to accommodate the evaluation of additional vendor products. As the type of CCD was previously discussed, let us begin our examination of digital camera features with the second entry, Focus Area, in the table.

Some digital cameras have a limited focus area, typically 10 or 20 feet to infinity. This type of camera precludes its use for close-up photographs and is obviously not an important consideration if you only anticipate acquiring images based on a distance of ten feet or more. However, if you need to use close-up photography, it is important to note that many cameras are capable of supporting optional lenses, either manufactured by the camera vendor or a third party that supports close-up photography, wide angle photography and other focus areas you may require. Unfortunately, the cost of some lenses may be significantly more than equivalent lenses for popular 35 mm cameras due to the limited number of digital cameras presently being manufactured. Thus, you may wish to consider your lens requirements, including adding their cost to the cost of a basic camera to obtain a more realistic indication of the total camera cost for each vendor being considered.

The images captured by a digital camera must be moved from the camera to a computer for processing. To do so you can either cable the camera to the computer, or transfer a PC Card used by some cameras. The manner in which cabling is accomplished is governed by the camera interface. The most common digital camera interface is a Small Computer Systems Interface (SCSI). A few cameras include a parallel interface, which enables the camera to be cabled to a computer's parallel printer port.

Concerning the use of PC Cards, although some cameras use the thicker Type III memory cards, most low-cost digital cameras use Type I or Type II cards that have a lower storage capacity because the resolutions of those cameras are less than those of more expensive cameras. Although a Type I card can fit in a Type II slot, the thicker size of a Type III card precludes its use in Type I and Type II card slots. Thus, you may wish to consider the type of PC Card slots in your notebook or in a PC Card reader attached to your desktop computer in conjunction with the type of PC Card used by a digital camera.

The image resolution is a function of the size of the CCD array built into the camera. In addition to the previously discussed resolutions, two additional common image resolutions are 640 by 480 and 758 by 504. Both of these image resolutions are associated with low-cost digital cameras which, although capable of providing an image suitable for a quality display on a VGA monitor, when the image is printed it results in an image which is far from the quality of a photograph. However, if you are looking for a method to rapidly and easily add a picture of a building, work area, person or object to a Web page, the use of a lower-resolution digital camera can provide the mechanism to satisfy this requirement.

Popular digital cameras that have 640 by 480 resolution include the Apple Computer QuickTake 150, Casio QV-30, Chinon Electronic Still Camera (ESC), Epson Photo PC, and Fuji DS-220. Examples of 758 by 504 image resolution cameras are the Kodak DC 40 and the Kodak DC 50 Zoom camera.

Although the digital cameras mentioned in this section have significantly lower resolutions than other cameras mentioned in this section, their costs are significantly lower. For example, the retail price of the Apple QuickTake 150 is $650, and the Chinon ESC has a list price of $499. The Kodak DC 40 with a retail price of $679 provides a slightly better image resolution and requires the use of an SVGA compatible monitor to view the difference between images produced by each camera. If you do not have a 600 dpi resolution laser printer and simply want a reliable camera to capture images that can be directly used when preparing a visual database, to add images to a Web page, or to place images on brochures, then any one of those cameras may provide a viable solution to your requirement to import pictures into a computer without a scanner.

As noted in Chapter 2, the color depth refers to the number of bits used to store the color of each pixel. Almost all low-cost digital cameras have a color depth of 24 bits; however, a few cameras permit users to record images as black and white using a gray scale color depth of either 4 or 8 bits. Some more expensive cameras use 12 bits/color, resulting in a 36-bit color depth that provides more detail for shadow areas and highlights. However, such detail is usually lost when this type of image is converted for use on a Web page or in a wordprocessing document.

The amount of camera RAM memory, as well as the image resolution and color depth, govern the number of images that a digital camera can store. For example, the Apple Computer QuickTake 150 includes 1 Mbytes of built-in RAM which enables the camera to store sixteen 640 by 480 images. In comparison, the Kodak DC 40 contains 4 Mbytes of RAM which enables this digital camera to store up to forty-eight 758 by 504 resolution images. In comparison, some professional digital cameras, such as the Kodak DCS 460 that has a resolution of 2036 by 3060 pixels and stores RGB images as 12 bits per color, requires 18 Mbytes to store an image. This resulted in the use of a Type III PC Card for this camera, which enables multiple images to be stored on the 170 Mbyte capacity of the card.

To provide additional image storage capacity, many low-cost digital cameras support two image quality modes. For example, the Chinon ES-3000 and Epson Photo PC both support 640 by 480 high-resolution and 320 by 240 low-resolution modes. This enables such cameras to store twice as many photographs when used in their low-resolution mode of operation.

A relatively new feature being added to digital cameras is internal image compression. Through the use of internal image compression, images are stored in compressed form in RAM. This enables more images to be stored before either needing to download images from the camera into a computer or writing over a previously stored image.

The primary method of internal image compression being incorporated into digital cameras is based on the Joint Photography Experts Group (JPEG) standard. That standard supports lossless compression as well as lossy compression. Concerning the latter, a user can specify the percentage of loss via a quality value. As it is not practical to provide a numeric keypad on a digital camera, vendors that incorporate JPEG-based compression commonly limit its use to lossless and 25% lossy via a switch setting. The selection of a 25% lossy compression can reduce the image's data storage requirement by 50% or more, while resulting in an image which visually cannot be distinguished from a non-compressed image.

The ability to use images captured by a digital camera is dependent on the software bundled with the camera. That software, designed to operate on a specific computer platform, controls the retrieval and conversion of images previously captured and stored in the camera. As many camera manufacturers use proprietary methods to internally store data, a program designed to retrieve data from a specific camera is commonly limited to use with that camera and cannot be used with other cameras. However, their bundled software may be applicable to use with images created by any camera once those images are transferred onto a computer. This is because most retrieval software permits a user to specify the storage of the image on a computer using a TIF, GIF, JPEG or another image storage standard. Once the image has been stored using a standardized image format, other programs, including image editing software that is either bundled with the camera or obtained separately, can be used. Some cameras, such as the Apple Computer QuickTake 150, include image manipulating and cataloging software which, if purchased separately, could result in an expenditure equal to a high percentage of the camera's retail price which includes the bundled software. Thus, if you are comparing competitive digital cameras, you may wish to adjust the cost of cameras that do not include bundled programs that you require by the cost of those programs if purchased separately.

Most digital cameras carry a one-year parts and 30 to 90 day labor warranty. When purchased through a retail store, it is often possible to obtain a warranty extension for a nominal fee. When considering a warranty extension, it is important to recognize that, although its name implies that it represents coverage beyond what the manufacturer provides, in actuality it does not. For example, a one-year parts and labor warranty purchased for a camera that comes with a one-year parts and 90 day labor manufacturer's warranty only adds 270 days of labor to the basic warranty.

The scanner

Although you can purchase CD-ROM based picture art containing hundreds of photographs, as well as diskettes containing half a dozen images focused on different topics, often you will prefer to customize your computer-generated documents such as Web pages and wordprocessing generated reports, or presentations by including images from photographs, magazine articles, forms and other documents. The key to obtaining this capability is the use of a scanner, a device which you can use to capture an image and convert it into a sequence of binary values that can be stored as a file on your computer. However, before discussing the use of a scanner, a word of caution concerning copyright material is needed. Unless you took the photograph, control the rights to a document or created a drawing that you wish to scan, you must obtain permission to use the item you want to scan.

As with other computer peripheral devices, there are many types of scanner, and their operation and functionality can significantly differ between different vendor products as well as products within one vendor's product line. Thus, the purpose of this section is to provide readers with information required to understand the functionality and capability of scanners. It also uses the presented information as a foundation for evaluating scanners to obtain an appropriate product to satisfy your specific imaging and computing requirements. To accomplish this goal this section includes an evaluation worksheet that enables you to match your specific imaging requirements against the functionality of different vendor products.

Functionality

The functionality of a scanner as well as its ability to be used with a specific type of computer is dependent on six general characteristics. Those characteristics include the type of scanner, its sensing technology, its pixel depth, the number of scanning passes it performs, its physical interface and the functionality of software provided by the scanner manufacturer. In this section we will examine each of those scanner characteristics to obtain an appreciation for how they govern the functionality of this data input device and govern its ability to be used with different types of personal computer.

Types of scanner

There are five basic types of scanner. Those types include flatbed, sheetfed, drum, slide and handheld devices.

A flatbed scanner is also commonly referred to as a desktop scanner. This is due to its physical characteristics, which resemble a rectangular box with a cover that lifts in a similar way to a desktop copier. This type of scanner is normally limited to digitizing letter or legal-size documents or portions of large blueprints or drawings that exceed the physical dimensions of its scanning area. Many flatbed scanners can be equipped with an optional transparency adapter that enables the device to scan $4'' \times 5''$ to $8'' \times 10''$ transparencies, overheads and 35 mm slides.

The resolution of flatbed or desktop scanners ranges from 300×300 pixels per inch (ppi) to 600×600. The higher-resolution devices are usually sufficient for desktop publishing; however, some graphic arts users who create sophisticated and detailed layouts that require higher resolution will probably require the use of a drum scanner. The retail cost of good-quality flatbed scanners range between \$500 and \$1000.

From a physical perspective a sheetfed scanner resembles a flatbed scanner in that both devices are rectangular. Although a flatbed scanner has a cover that lifts up to enable the insertion of a document, the sheetfed scanner feeds a single sheet at a time through the device, where it is scanned. This design constraint limits the use of the sheetfed scanner to individual pages, making it unsuitable for scanning pages in bound books and magazines, large documents beyond the capacity of the sheet feed area, posters and other documents. In fact, many sheetfed scanners are limited to scanning text documents. Thus, the primary use of a sheetfed scanner is to scan a large number of batched pages or for optical character recognition applications, where special software used on a computer connected to the scanner interprets the scanned dot pattern and converts recognized characters to their appropriate eight-bit byte code.

Although the basic hardware cost of a sheetfed scanner is normally less than that of a flatbed scanner, the total system cost can exceed that of a flatbed scanner by a considerable margin. This is because the OCR software cost can range from being included with the hardware for elementary systems, to costing \$10 000 or more for sophisticated programs that minimize the character recognition error rate to under one character in a million.

A drum scanner captures image data at a higher resolution than that obtained with flatbed and sheetfed scanners. Owing to the rotation of the drum across a scanning area, this type of scanner is suitable to digitize large blueprints and drawings that cannot be scanned as an entity by flatbed and sheetfed scanners.

From a physical perspective a drum scanner is significantly larger than a flatbed scanner and may require professional training to operate correctly. From a cost perspective the higher resolution, usually up to 1200×1200 ppi, and ability to move a document around a rotating drum results in a more complex mechanism whose cost can be upwards of \$30 000. Owing to the high cost of drum scanners it is common for graphic arts users to perform their layout work using a flatbed scanner and use a drum scanner at a service bureau to obtain the required digitization.

A slide scanner is limited to digitizing transparencies, typically 35 mm slides. Although this type of scanner can produce as high a resolution as a drum scanner, it does so for a very small scanning area and its resulting image normally has to be enlarged to be used in documents. In addition, this type of scanner is limited to working with transparencies and cannot be used to scan reflective documents, such as photographs and line art drawings. The cost of a slide scanner ranges between $500 and $1000 and many devices are sold as an option for use with a flatbed scanner.

As its name implies, a handheld scanner is a device that you place in your hand and move across a document to scan its image. Handheld scanners were originally limited to gray scale digitization, but within the past few years vendors have introduced color versions.

Although most handheld scanners have a relatively low resolution of 300×300 ppi, some newer models can be obtained with a resolution of up to 800×800 ppi. Although the cost of handheld scanners are typically under $200 for gray scale devices and under $350 for color models, their capability is highly dependent on the operator. This is because the sampling speed and the straightness of the scan depend on the person holding the scanner.

Sensing technology

Scanners incorporate one of two sensing technologies which provides them with the ability to recognize images: charge coupled devices (CCDs) or photomultiplier tubes (PMTs).

In a CCD scanning process a light source is moved to illuminate a line on an image. An array in a row on a single silicon chip converts the reflected light into an analog voltage. That voltage is then converted through the use of an analog to digital converter into a digital value. That is, darker areas absorb light and reflect less, resulting in a lower measured voltage, and lighter areas reflect light and have a higher reflection resulting in a higher measured voltage. As the cost of CCDs is relatively expensive, a mirror is used to direct consecutive light scans to a static array of CCDs, minimizing the number of charge coupled devices required to sense light reflections.

Flatbed, sheetfed and handheld scanners are commonly based on the use of CCD technology. The resolution obtainable through the use of a CCD-based scanner depends on the number of CCD elements used in the CCD array and the length of the scan line. For example, consider a CCD-based flatbed scanner with 6800 elements in its CCD array. If the scanner is capable of scanning documents up to $8\frac{1}{2}$ inches wide, then its horizontal scanning rate is 6800/8.5 or 800 pixels per inch (ppi), a term also commonly referred to as dots per inch or dpi.

The actual conversion of the analog voltage produced by the elements in the CCD array into a digital sequence of bits occurs from the use of an analogue-to-digital (A/D) converter. The number of bits generated by each A/D converter for each sample determines the number of gray levels or color resolution of the scanner. For example, an 8-bit A/D converter provides 8 bits per sample that can represent 2^8 or 256 gray levels. By using a three-pass process, three 8-bit samples can represent 2^{24} or approximately 16.7 million different colors per pixel.

A second type of scanning technology is based on the use of photomultiplier tubes, and it is primarily incorporated into drum scanners. The PMT tubes are glass vacuum tubes, similar to the type of tubes once commonly used in radios and televisions. A light-sensitive metal fastened inside the tube, known as an emitter, generates electrons which traverse a small space in the vacuum and collide with a metal detector. This collision results in the generation of an electric current. Similarly to a CCD, the current generated by the PMT varies based on the brightness of a reflected light source.

As you might suspect, within a drum scanner the object to be scanned is mounted on a rotating drum. A series of mirrors position the light source in the horizontal plane to a specific location, and the rotation of the drum enables a continuous strip to be illuminated and have the reflection focused onto a series of PMTs without moving the light source. Thus, the light source only needs to move in the horizontal plane to illuminate the vertical and horizontal planes of the document mounted on the drum.

The use of PMTs results in a very sharp image because they have a higher sensitivity to light than CCDs. In addition, their voltage range exceeds the output of CCDs, which enables the A/D converter to have each resulting bit represent a greater level of detail. This enables a PMT-based scanner to capture image details often passed by a CCD-based scanner. However, this capability is not without cost, as drum-based scanners using PMTs can cost 20 to 30 times the cost of flatbed CCD based scanners.

Pixel depth and scanning passes

The capability of a scanner to distinguish variations in the reflection of light from an illuminated image is dependent on the amount of information that can be stored for each pixel, or pixel depth.

Pixel depth depends on the analog-to-digital converter as well as the type of image scanned. Concerning the former, a 4-bit A/D converter provides 16 gray levels whereas an 8-bit A/D converter provides a 256 gray level capability. Concerning the latter, a line art drawing only requires pixel depth of one bit to fully capture the image as the pixel is either on (white) or off (black) at any point in the image. In comparison, a black and white photograph with more than 16 gray levels could not be fully captured through the use of a scanner with a 4-bit A/D converter.

The pixel depth of color images is typically expanded to 24 bits through the use of eight bits for red, green and blue primary color components. Those components are combined electronically to provide approximately a 16.7 million color support per pixel. Although some scanners now support up to 32 bits per pixel, the colors supported by the use of 24 bits represents the most that can be visually identified by the human eye.

Color scanners use either a one-pass or three-pass scanning technique to separate and then recombine the three primary colors. In a one-pass scanning technique three light sources are turned on and off for each scan line. This technique uses three light exposures to captures the line image in one pass and can be implemented using colored light sources or via the use of a white lamp with colored filters.

In a three-pass scanning technique the line is first scanned with one color lamp illuminated. Next, the process is repeated for each of the remaining two colors. As a slight variation in the stability of a trio of scans can 'fuzz' a color picture line, a three-pass scanner is prone to slight vibrations as well as an inability to produce perfect mirrored scans as the scanner ages. Although it takes longer to make three passes than one, the overall scanning time is less, because this method does not require the processing time required to combine three 8-bit samples. Thus, a three-pass scanner is more suitable for real-time or near real time applications, such as capturing and displaying a series of images through the use of a scanner connected to a computer whose display post in turn is connected to a projection screen.

Interface

The physical interface of a scanner governs its ability to be used with different types of computer systems. Most handheld scanners use either a serial or parallel interface, permitting them to be cabled directly to a computer's built-in serial or parallel port. Flatbed, sheetfed and drum scanners are primarily manufactured with an SCSI interface and require the installation of an SCSI adapter card in your computer. As most laptop computers and almost all notebooks lack a system expansion port, some vendors offer a parallel port option that enables their use with devices that cannot support an SCSI interface. In addition, the use of a parallel printer port interface may be required if your computer already has an SCSI board supporting an internal CD-ROM. This is because many internal CD-ROMs cannot support SCSI daisy chaining, and SCSI adapter boards from different vendors are notoriously incompatible, causing peripheral incompatibility or even the computer to lockup.

Software

Both the operating system supported and scanner functionality are key areas of concern with respect to software. The operation system support governs whether the scanner will work on a Macintosh, on a PC platform or under different versions of Windows or OS/2. Concerning scanner functionality, software is the key to whether or not the scanner will perform optical character recognition, its ability to automatically recognize where color images and halftones are positioned on a page and adjust its scan to subsections of a page and preview scanned images and provide you with the ability to adjust the scanning process. As the results of most image scanning operations are designed for use within an application, you may wish to determine the compatibility of a scanner's software with a standard known as TWAIN. TWAIN is an industry standard interface which permits image data to be acquired from an external source without having to exit your current application. If your scanner software is TWAIN-compliant, you can use it directly with over 100 software products.

Evaluation characteristics

Now that we have reviewed the general types and characteristics of scanners we can use our previously acquired knowledge as a foundation for evaluating this category of computer peripheral device. Table 4.3 contains an evaluation worksheet you can use to

Table 4.3 Scanner evaluation worksheet

Function/Feature	Requirement	Vendor A	Vendor B
Type of scanner			
Flatbed	_____	_____	_____
Sheetfed	_____	_____	_____
Drum	_____	_____	_____
Slide	_____	_____	_____
Handheld			
Sensing technology			
Charge coupled device	_____	_____	_____
Photomultiplier tube	_____	_____	_____
Pixel depth			
Gray scale levels			
16	_____	_____	_____
256	_____	_____	_____
Color support level (bits)			
8	_____	_____	_____
16	_____	_____	_____
24	_____	_____	_____
Scanning process			
One	_____	_____	_____
Three	_____	_____	_____
Interface			
Serial	_____	_____	_____
Parallel	_____	_____	_____
SCSI	_____	_____	_____
Software			
OS supports			
Macintosh	_____	_____	_____
DOS	_____	_____	_____
Windows	_____	_____	_____
Functionality			
OCR capability	_____	_____	_____
TWAIN standard support	_____	_____	_____
General features			
Cost	_____	_____	_____
Manuals	_____	_____	_____
Training support	_____	_____	_____
Warranty	_____	_____	_____

compare and contrast your requirements to a comprehensive list of scanner features. In addition, the columns labeled Vendor A and Vendor B provide you with the ability to evaluate vendor products against your requirements. By adding additional columns to the right of Table 4.3 you can evaluate additional scanners if you wish to.

Operational considerations

The selection of an appropriate scanner to satisfy your imaging requirements primarily depends on the type of documents that you will scan, the application or applications that you intend to place images into and the computer platform that you intend to use. The type of documents to be scanned will narrow the type of scanner most suitable for use. The application or applications with which you intend to use images will govern the pixel depth, including gray scale or color levels required. The type of computer platform will govern the scanner interface and software that must operate on the computer to control scanner operations. By carefully considering your imaging requirements and matching those requirements against the functions and features listed in Table 4.3 you can develop a core set of scanner specifications. Those specifications can then be used as a basis to evaluate different vendor products to obtain a scanner that best meets your requirements. Concerning those requirements, it is important to note that the resolution of a scanner should be considered with respect to the type of images or documents that you anticipate scanning, as well as the device that the scanned image will be displayed on. If you anticipate scanning a newspaper, most images are printed at 85 lines per inch. As it is best to scan images at a resolution of 1.5 to 2 times the resolution of a document, you would set your scanner's resolution to 170 ppi.

When scanning photographs for use on Web pages or for display on monitors, it is important to note that most computer screens use a resolution of 72 ppi. Thus, scanning a photograph at a higher resolution will result in a much larger file than actually necessary. For example, consider a 3×5 inch color photograph scanned at 300 ppi using a color depth of 8 bits per pixel. The resulting file would require 1.35 Mbytes of storage and, if placed on a Web page whose server is connected to the Internet via a 64 Kbps digital line, would require approximately 169 seconds (300×300 ppi $\times 15$ inches sq. $\times 8$ bits/byte $\div 64\,000$ bps) or almost 3 minutes to download. In comparison, if the photograph was scanned at 72 ppi, the resulting file would require 77 760 bytes of storage and the time required to download that file would be reduced to under 10 seconds!

Screen capture

An increasing source of images results from the use of screen capture programs which are used to store all or a selected portion of a monitor's display. Although the primary use of screen capture programs is in the area of education and publishing, because it enables authors to easily incorporate illustrations of computer operations with text, such programs are also used to capture drawings generated by application programs that produce vector image files that may not be directly suitable for use in other applications, such as being placed on a Web page. This is because most browsers are limited to supporting GIF and JPEG images and require plug-in viewers to obtain the ability to view other types of images.

Windows clipboard

Although not commonly recognized as such, the Windows clipboard provides a screen capture capability. When you cut or copy information from an application or copy an image of the active window by pressing the ALT + Print Screen key combination or capture the entire window by pressing the Print Screen key, the information is first sent to the clipboard. Included in Windows is a ClipBook Viewer which provides you with the ability to save the contents of the clipboard, view previously stored clipboard contents, and edit their contents.

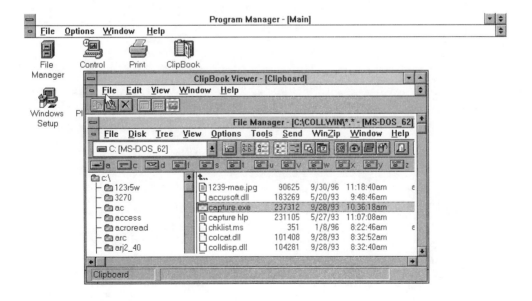

Figure 4.3 Using the Windows ClipBook Viewer to view the contents of the clipboard

Figure 4.3 illustrates the use of the ClipBook Viewer on the author's computer that shows a portion of a previously 'printed' screen that resides in the clipboard. The Windows ClipBook Viewer saves files in its default mode using the file extension .CLP. CLP is a Windows Bitmap image file format that is supported by several, but not all, image manipulation programs. As many programs support multiple image file formats, you can save an image 'printed' to the screen and then convert it into another format if your application does not support the CLP format.

Commercial programs

Although the clipboard provides a mechanism to capture images, its use is limited to capturing either the entire screen or the active window. In comparison, many commercial and shareware screen capture programs include a large number of features that allow you to save selected portions of a screen and store the capture image in a variety of different file formats. As the focus of this

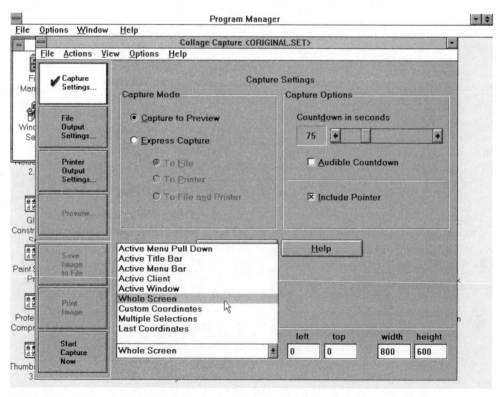

Figure 4.4   Using the Collage Capture Capture Settings button

section is on screen capture programs, Let us briefly examine the use of one program that has already been used a few times in this book. That program is the Collage Capture program which, with Collage Image Manager, is sold under the name Collage Complete by Inner Media, Inc. Of Hollis, NH.

Figure 4.4 illustrates the Collage Capture screen with its Capture Settings option button and its Cropping Area menu selected. Through the use of the Capture Settings button you can specify the time in seconds during which the program resides in the background prior to rising to capture the screen image. The Cropping Area menu shown in the lower left portion of Figure 4.4 allows you to select the portion of the screen to be captured and can be changed after the initial selection. Figure 4.5 illustrates the selection of the File Output Settings button. The selection of this button allows you to specify the file and the location where the capture image will be stored, as well as its brightness, scaling, cropping area and style. Concerning the latter, the program refers to the format, use of compression, color depth, and pattern

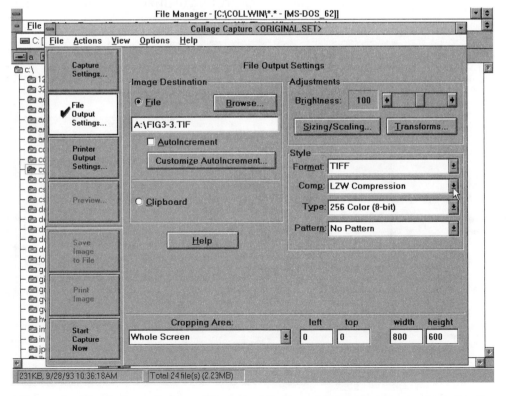

Figure 4.5 The Collage Capture File Output Settings button enables you to define the image format and format options used to store a captured image

selections as style. Here the term pattern represents the dithering method if you either selected black and white as the Type (color depth) or area capturing an 8 or 24-bit color or gray scale screen to an image of fewer colors or gray scale.

A third button option of the Collage Capture Program that deserves mention is its Preview button. The use of this button, which is shown in Figure 4.6, enables you to view the captured screen as well as use the Cropping Area menu to select a different portion of the screen. Collage Capture, as well as similar screen capture programs, provides you with the ability to capture to disk or print all or selected portions of a screen image. Once captured, if necessary, you can convert the image file format into a more appropriate file format and use it in an almost infinite number of applications.

Screen grabbers

This author uses the term screen grabber to refer to a variety of products developed to capture and digitize pictures from camcorders,

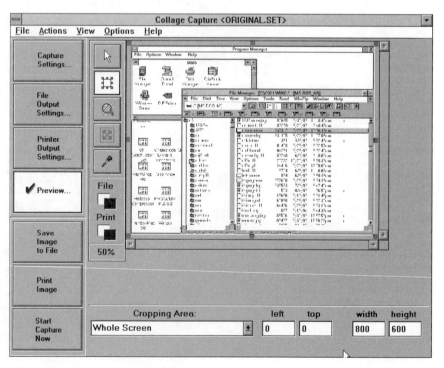

Figure 4.6 Using the Collage Capture preview button to view the image to be captured

VCRs or television. Some screen grabber products require the installation of a special adapter card into your computer. Other screen grabbers are manufactured as stand-alone devices similar in size to a VCR tape and include video in and out ports and a serial or parallel port for connection to your computer. Through the use of software supplied with the screen grabber, you obtain the ability to save grabbed video in a variety of image formats to include different compressed formats.

4.2 COMMUNICATIONS-BASED IMAGE ACQUISITION TECHNIQUES

There are three basic methods used to acquire images via communications: bulletin board access, LAN-based access, and via a communications link to the Internet. The latter represents a catch all technique that encompasses conventional file transfer via FTP and Web browsers, as well as more exotic methods including the attachment of encoded binary files to e-mail. Concerning the latter technique, it is also possible to use e-mail to transfer attached files on an intra-company basis. As the transfer of binary images as e-mail attachments is incompatible with 7-bit ASCII supported e-mail gateways, a conversion must occur to enable 8-bit binary files to flow through such gateways. This conversion process is described later in this book. Thus, we will defer a detailed discussion of the use of binary attached files with e-mail.

Before discussing image acquisition techniques via communications, a word of caution is in order concerning copyrights. Even though an image may reside on a public access bulletin board or Internet file server, this fact does not by itself denote that the image is not copyrighted nor does it imply permission for commercial use. Perhaps the best policy to perform concerning the use of images is that when in doubt, don't!

Bulletin board access

Both commercial and public bulletin boards primarily store images in a compressed format, typically in a JPEG, TIF or GIF format. Although this precludes the ability to enhance file transfers by running a compression utility on the bulletin board prior to performing the file transfer, depending on the number and type of images to be transferred and the bulletin board software used, you may have the ability to enhance your data transfer operation. For example, if you require the transfer of two or more files, you should use a file transfer protocol that supports the transfer of multiple files, such as ZMODEM or YMODEM, instead of XMODEM that is limited to the transfer of one file at a time. As ZMODEM is a full-duplex file transfer protocol, whereas YMODEM and XMODEM are half-duplex file transfer protocols, this means that the use of the former can result in faster file transfers. As each protocol requires one or more transmitted data blocks to be acknowledged prior to the next block being transmitted, the ability to transmit an acknowledgment in one direction while a block flows in the opposite direction enhances data throughput. This means that ZMODEM should be your file transfer protocol of preference, followed by YMODEM and ZMODEM. As the cost of a long-distance telephone call is commonly based on the duration of the call, any technique that you use to reduce transmission time also reduces the cost of transmission.

LAN-based access

Two of the more common evolving LAN-based applications are the establishment of image servers and the use of CD-ROM jukeboxes attached to file servers to provide workstation users with the ability to access centrally located clip art, photographs, drawings and other types of image. As CD-ROMs have relatively long access times in comparison to hard drives, you should consider transferring the contents of the former to use the latter to significantly enhance network access to stored images.

Owing to the significant decrease in the cost of hard drives over the past few years while their storage capacities have dramatically increased, it makes little sense to use a $200 CD-ROM to store a 600 Mbyte collection of images when for approximately the same price you can purchase a 1.2 Gbyte hard drive. Thus, unless there is a compelling reason to use CD-ROMs you will be able to enhance network access and obtain an improvement in the ability to store more data for the same storage cost by switching to hard drives.

Internet access

There are two primary methods that you can use to transfer images via the Internet: the File Transfer Protocol (FTP) and Web browsers.

FTP

FTP predated the World Wide Web by approximately 30 years. Although Web browsers garner the majority of Internet publicity, most Web server platforms, such as Microsoft's Windows NT Server, include an FTP server capability. This capability enables many users to employ an FTP client to download files far more easily than through the use of a Web browser.

Figure 4.7 illustrates the use of the Chameleon FTP client program developed by NetManage to access an FTP server. The Chameleon FTP client window is subdivided, with the left portion indicating local directories and files and the right portion doing the same for the remote host. The buttons in the middle of the screen which point to the left and right perform the indicated operation on the directory or file selected. In Figure 4.7, the highlighted file INDEX.HTM on the remote host will be copied to the file DUMMY.HTM on the local host by clicking on the button labelled COPY that points to the left. By providing users with the ability to easily view the contents of directories and transfer files,

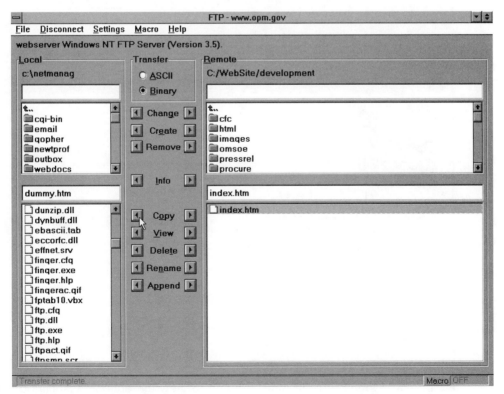

Figure 4.7 Using the Chameleon FTP client to download a file from an FTP server

the use of a GUI-based FTP can facilitate the transfer of files many times faster and easier than the use of a Web browser.

As many Internet users have either unlimited usage accounts or else access the Internet via a corporate LAN that has a permanent connection, the primary gain obtained from reducing transmission time involves productivity instead of cost. By carefully examining the contents of FTP sites, you may be able to reduce your transmission time and obtain the ability to perform other functions earlier than otherwise possible, enhancing your productivity. To accomplish this, you should examine the FTP site to determine if the images that you seek are stored in different file formats, and, if so, download the most appropriate format. For example, if an image is stored in GIF and JPG formats and JPG is acceptable, you may be able to decrease your download time by a factor of ten or more. Similarly, if the FTP server stores both individual images and an archive of images as a ZIP or similar file, if you require each image you should download the ZIP file instead of performing a series of individual downloads.

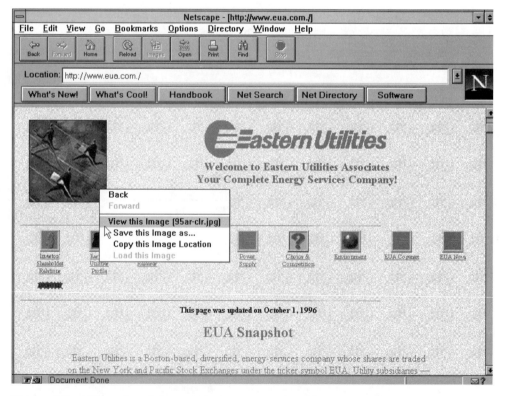

Figure 4.8 Using the right mouse button to view the options associated with an unlinked image

Web browser

In its default mode of operation most Web browsers load and display images associated with a specified page. Once you display a page with one or more graphics, most browsers support the generation of pop-up menus when the right mouse button is clicked. Those pop-up menus list items that commonly represent shortcuts for initiating several browser commands.

When you place your cursor over an image and click the right mouse button, the pop-up menu will display menu items that primarily refer to the image file used to generate the image. As an example of the use of the right mouse button, consider Figures 4.8 and 4.9 which illustrate pop-up menus associated with two images on the same Web page.

In Figure 4.8 the image on which the right mouse button was clicked has no link associated with it that limits the options in the menu. The View option, which is highlighted in Figure 4.8, indicates the name of the file on which the image is stored on the

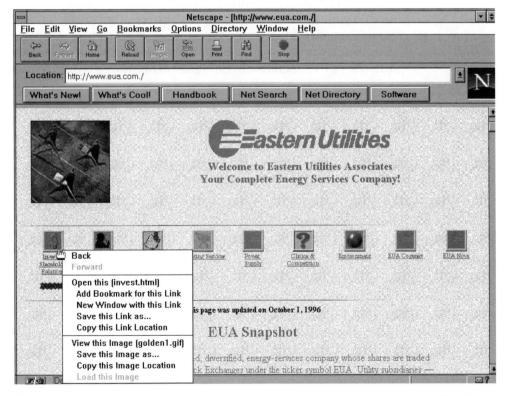

Figure 4.9 Using the right mouse button to view the options associated with a linked image.

distant server whose home page was accessed and displayed. By selecting the View option, the image icon is replaced by the corresponding image on a separate screen. The Save option enables you to save the specified image to disk, and the Copy option results in the location of the specified image in terms of its Uniform Resource Locator (URL) being copied into the Window's Clipboard. Through the selection of the Save option, Windows will display a dialog box that enables you to save the image with the current file name or a different name. Thus, a simple right click paves the way to copy an image from a Web page to your computer.

Figure 4.9 illustrates the pop-up menu resulting from pressing the right mouse button when it was positioned on a linked image. In examining the pop-up menu shown in Figure 4.9, note that the lower portion of the menu is equivalent to the second portion of the pop-up menu when it is placed on a non-linked image. For both menus, selecting the 'Save this Image as' option provides you with the ability to save the image in its present file format. In Figure 4.8 this would result in saving the image as a JPG file, whereas in Figure 4.9 the image would be saved a GIF file. Once the desired image or images have been saved, you can use an image manipulation program to convert the image into a more desirable format, if required. Thus, the use of a Web browser facilitates the extraction of images used on Web pages, whereas the use of FTP facilitates the retrieval of images stored on a server.

IMAGES AND THE LAN

When images are used on an individual computer, any problems associated with their use are localized. In comparison, when images are placed on a local area network the use of such images by one network user can adversely affect other network users. Owing to the impact of LAN-based images on network performance, we will examine the effect of images on LAN bandwidth as well as techniques that can be used to minimize their effect. As LANs are the foundation for the construction of corporate intranets, our examination of LAN-based images is also appropriate to understand the effect of images on corporate intranets and techniques that you can consider to improve intranet performance.

5.1 THE IMPACT OF IMAGES

The provision of information concerning the impact of images on LAN bandwidth is facilitated by having a few examples to work with. In writing this chapter this author turned to his photograph archive to select two pictures that could be used to represent the use of images on a local area network. The first photograph, which is shown in Figure 5.1, is of the author's daughter and was gray scale scanned using a resolution of 100 by 100 dpi. The resulting image, which is 472 by 324 pixels in size, was stored using a TIF file format without compression, and it required 153 170 bytes of storage. This photograph could represent a photo ID incorporated into a personnel database. The second photograph taken from the author's photograph archive is that of a most interesting apartment building in Tel Aviv. This building, which is illustrated in Figure 5.2, was gray scale scanned using a resolution of 200 by

Figure 5.1 A scanned photograph of the author's daughter that will be used to illustrate the effect of photo ID's added to a LAN-based personnel database or similar application

200 dpi. The resulting image, which is 632 by 784 pixels in size, was also stored using the TIF file format without compression, requiring 495 694 bytes of storage. This image could conceivably represent an image incorporated into a real estate database. Now that we have two images to work with, let us discuss the potential effect of their transmission on different types of popular LANs. Doing so will provide us with an understanding of how images impact LAN bandwidth.

Shared media constraints

With the exception of ATM, other types of LAN are shared media networks. This means that regardless of the type of LAN, such as Ethernet, Fast Ethernet or Token Ring, access to the network is shared by many users. This also means that on the average a network with an operating rate of Kbps with n stations provides K/n bps to each user.

To illustrate this facet of shared media networks, consider Figure 5.3 which illustrates a basic Ethernet 10BASE-T network hub supporting seven workstations and one file server for a total of eight network devices. Although the network operates at

Figure 5.2 A scanned photograph of an unusual building in Tel Aviv that will be used to illustrate the effect of images added to a LAN-based real estate application

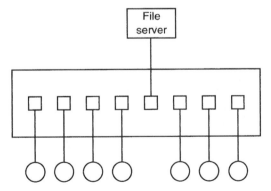

Figure 5.3 Basic Ethernet 10BASE-T network. Although data transmission on a 10BASE-T network occurs at 10 Mbps, over a period of time each station obtains only a fraction of the total transmission capacity based on the number of stations connected to the network

10 Mbps, only one network device at a time can gain access to the network. In fact, if two or more devices attempt to transmit at the same time, a collision will occur, resulting in each station's adapter card initiating a random exponential backoff algorithm which generates a time delay value used by the station prior to its attempting to retransmit. This explains why the Ethernet LAN access protocol is referred to as Carrier Sense Multiple Access/ Collision Detection (CSMA/CD) where the carrier sense term refers to the method by which Ethernet adapters listen to network activity to determine whether or not transmission is occurring. Even though a network adapter card may not sense the presence of a carrier on the network, this fact by itself does not mean that the adapter can transmit a frame of information without a collision occurring. This is because a signal could have recently been placed on the network by another network adapter card that has not yet reached the first adapter card. As the first adapter card places data on the network, a collision would then occur.

The effect of shared media

To illustrate the effect of shared bandwidth on the transfer of network-based images, let us consider the transfer of Figures 5.1 and 5.2 from a server to a workstation. If it was possible to obtain the full 10 Mbps bandwidth of the LAN, the transfer of Figure 5.1 would require 153 170 bytes × 8 bits/byte ÷ 10 Mbps, or approximately 0.1225 seconds. In comparison, the transfer of Figure 5.2 would require 495 694 bytes × 8 bits/byte ÷ 10 Mbps, or 0.3966 seconds. It should be noted that both photographs were scanned using a 256 gray level. If a 24-bit true color format was used to store each scanned photograph, their storage requirements using non-compressed TIF would treble. This means that the time required to download each image would also treble, resulting in a true color version of Figure 5.1 requiring 0.3675 seconds to download, whereas a true color version of Figure 5.2 would require 1.1898 seconds to download.

If we discard the effect of collisions, over a period of time the simple eight-port Ethernet hub-based network illustrated in Figure 5.3 would provide each network device with an average 10 Mbps/8 or 1.25 Mbps of bandwidth. If Figure 5.1 was placed on the file server, the average time required to download that image would lengthen to 153 170 bytes × 8 bits/byte ÷ 1.25 Mbps, or approximately 0.98 seconds when you consider the fact that other network users also contend for network bandwidth. If a true color version of Figure 5.1 were stored on the file server, the time required to download a non-compressed TIF file could treble to approximately 3 seconds, reaching a delay where patience is required from a network user as the delay is very perceivable.

If we turn our attention to Figure 5.2, its download time when stored as a 256 level gray scale image on a shared media network would require 495 694 bytes × 8 bits/byte ÷ 1.25 Mbps, or approximately 3.2 seconds. If stored as a true color image, the download time would triple to 9.6 seconds, a delay that becomes very pronounced and begins to tax the patience of even a very tolerant person.

Considering network efficiency

Unfortunately, due to several reasons, the preceding computations represent a best-case scenario. First and foremost, no LAN is 100% efficient and Ethernet is no exception. Owing to the CSMA/CD protocol and the effect of collisions causing stations to delay subsequent transmissions, most 10 BASE-T Ethernet LANs provide a maximum transmission capability between 5 and 6 Mbps. Secondly, when an image is transported on a LAN, each frame has a degree of overhead. A possible third reason why the preceding computations are not realistic is the fact that most LANs have a significant number of stations beyond the eight shown connected to a single hub illustrated in Figure 5.3.

To illustrate the potential effect on the transmission of the two previously discussed images, let us assume that the LAN supports 72 users and provides and an average information transfer capability of 6 Mbps due to the effect of network collisions and random exponential backoff delays. Then, instead of dividing 10 Mbps by 72 to obtain the average LAN bandwidth per workstation, we would divide 6 Mbps by 72, because the information transfer capability of a network is a more realistic metric to use. Doing so, we would obtain an average transmission rate per LAN station of 83 333 bps (6 Mbps/72), which means that when other network users are performing network-related operations, the downloading of the first photograph scanned as a 256 gray level image, which was illustrated in Figure 5.1, would require 153 170 bytes × 8 bits/byte ÷ 83 333 bps or 14.7 seconds. If the image were scanned using a true color format, its storage requirements and transmission time would both treble. Thus, its downloading would expand to 44.1 seconds. Now we are talking about patience being a virtue! Similarly, the downloading of the second image which was illustrated in Figure 5.2 would require 495 694 bytes × 8 bits/byte ÷ 83 333 bps, or almost 48 seconds when stored as a 256 level gray scale image. If scanned as a true color image, the transmission time would again treble, in this example expanding to 144 seconds or almost two and a half minutes. Now you have enough time to enjoy a cup of coffee as you wait to receive the requested image. Although the preceding computations indicate much more pronounced transmission delays, in actuality they represent a worst-case scenario that deserves a degree of elaboration.

Delay computation considerations

The computation of an average transmission rate per LAN station assumes that all stations are active participants on the network. In actuality, during the day some workstations may not even be turned on as their operators may be traveling and away from the office. Other workstations that are powered on and being used do not necessarily consume network bandwidth. For example, an employee may be typing a memorandum or developing an electronic spreadsheet using local programs instead of accessing LAN-based programs. Although this means that each network request on a 72 node Ethernet network will obtain between 6 Mbps and 83 333 bps of transmission capacity, the actual bandwidth obtained by a network user will depend on the number of station users performing network related operations. Thus, a more precise estimation process must consider the average number of users performing network-related operations instead of the total number of stations on the network. Although this metric will obviously vary from network to network based on user activity, most networks have between 3 and 5% of network stations actively performing network-related operations at any point in time. This means that for a 72-station network you can reasonably expect between 2.16 and 3.6 users to be actively performing network-related operations. By recomputing the image transmission times we would use an information transfer rate between 2.77 Mbps (6 Mbps/2.16) and 1.67 Mbps (6 Mbps/3.6). This means that a more realistic transmission time for the 256 level gray scale image shown in Figure 5.1 would be between 0.44 seconds (153 170 bytes ×8 bits/byte ÷ 2.77 Mbps) and 0.73 seconds (153 170 bytes ×8 bits/byte ÷ 1.67 Mbps). Similarly, for Figure 5.2 a more realistic transmission time for that gray scale image would be between 1.43 seconds (495 694 bytes × 8 bits/byte ÷ 2.77 Mbps) and 2.37 seconds (495 694 bytes × 8 bits/byte ÷ 1.67 Mbps). For true color representations of each image, their storage and transmission times would treble. Thus, the true color image of

Figure 5.1 would require between 1.32 and 2.19 seconds for downloading, wheras the true color image of Figure 5.2 would require between 4.29 and 7.11 seconds.

To provide a timing comparison, Table 5.1 indicates the transmission time required for a workstation to download each image under four different scenarios for both 256 level gray scale and true color versions of each image. The first column in Table 5.1 indicates the time required to download each image under ideal conditions with no other network users performing LAN-related activities. Under this situation, it is quite possible for the information transfer rate to approach 10 Mbps due to the absence of collisions. There is some overhead in each Ethernet frame due to the inclusion of source and destination address fields, a type field and a CRC field. In addition, Ethernet requires a time separation of 9.6 µs between frames which, when considered together with the overhead of each frame, results in a maximum information transfer rate of approximately 9.9 Mbps when one station retrieves a file from another network station without any other network users accessing the network. Thus, the first column in Table 5.1 represents a best-case situation.

The second column in Table 5.1 can be considered to represent the most pessimistic transmission time. This is because the transmission times in that column are based on all network users being active which, due to collisions, results in a usable bandwidth of 6 Mbps. As indicated in that column, the transmission time to download the 256 level gray scale image shown in Figure 5.1 would increase to 14.7 seconds, whereas the time required to download the 256 level gray scale image shown in Figure 5.2 would increase to 48 seconds. When we examine the time associated with downloading true color versions of the two

Table 5.1 Image transmission time comparison

	(Transmission time in seconds)			
	10 Mbps One active user	6 Mbps 72 active users	6 Mbps	
			3% active	5% active
256 level gray scale				
Figure 5.1	0.12	14.7	0.44	0.73
Figure 5.2	0.39	48.0	1.43	2.37
True color				
Figure 5.1	0.36	44.1	1.32	2.19
Figure 5.2	1.07	144.0	4.29	7.11

images, we note that the transfer times increase to 44.1 and 144 seconds. As indicated by the time entries in column 2, the time required to download images on a heavily utilized network can be considerable. The third and fourth columns in Table 5.1 illustrate what are probably much more realistic transmission times based on 3 and 5% of a 72-user network performing network operations at any one point in time.

As indicated by our series of computations, the determination of the time required by a LAN-based workstation to download an image depends on a number of variables and can have a considerable range of values. If we focus our attention on the most probable transmission times based on 3–5% of network users being active, we might be tempted to say 'so what?' After all, what is the effect of a 0.44 to 2.37 second delay associated with the transfer of gray scale images and a 1.32 to 7.11 second delay associated with the transfer of true color images on employee productivity, and why would you want to consider modifying images or altering your LAN configuration to reduce the transmission time required to download an image? The answer to this multi-part question lies in the fact that the use of images, like candy, can be addictive with one LAN-based image application followed by another, and another, and another. When this occurs, it becomes relatively easy for just one or two network users flipping through a visual database, such as a personnel file that includes pictures of employees, to substantially degrade LAN performance experienced by other network-based users. For example, assume that over a short period of time a real estate broker working with a client in the office 'flips' through a series of commercial real estate listings containing digitized photographs of properties similar in resolution and size to Figure 5.2. As each image flows on the network it will require between 1.43 and 2.37 (gray scale) or between 4.29 and 7.11 (true color) seconds to reach its destination based on 3 and 5% of the network users being active. If another network user initiates a file transfer or another relatively long client–server operation while the image transfer is in progress, both operations compete for network bandwidth for a relatively long period of time in comparison to the time required for short interactive query–response operations. This means that the probability of collisions occurring will substantially increase, with each collision delaying the occurrence of retransmissions due to the random exponential backoff algorithm used by Ethernet.

We can sum up the effect of a sequence of images being transmitted on a network by stating that they result in the generation of cumulative delays that feed on one another. That is, not only does the subsequent downloading of a second image delay the response to queries or file transfers initiated by other network users, but it also results in an increase in the time required to download the following images. Thus, while the periodic downloading of an image may have a minimal effect on other network activity as the use of images increases, it will have a more profound effect on all network users. Owing to this many hardware vendors suggest changing the LAN infrastructure to better accommodate the use of network-based images. Although you may eventually need to change your network infrastructure to accommodate the transmission of images, hardware-based modifications only represent one of several options that you can consider to more efficiently and effectively work with network-based images. Thus, in the next section in this chapter we will turn our attention to a variety of techniques that can enhance the efficiency of image-based applications on a network.

5.2 ENHANCING NETWORK-BASED IMAGE APPLICATIONS

Now that we have an appreciation for transmission delays that result from the transport of images on a LAN, we can turn our attention to methods that minimize their effect. In doing so we will first discuss and describe software-based techniques as their implementation can occur with a minimal amount of cost and effort. Although the use of a variety of software-based image modifications can substantially minimize the effect of image transmission, the quantity of images to be transported, the current level of network utilization or an increase in network activity can result in the necessity to perform a more costly modification to your network infrastructure based on the addition or modification of hardware. Thus, after we discuss and describe software-based techniques, we will turn our attention to hardware-based techniques.

Software-based techniques

Assuming that you have already acquired one or more images, you can enhance their use by adjusting them to the intended application. To do so you can consider using one or more image manipulation programs that provide you with the ability to adjust images to the application that they will be used by. Such adjustments can include cropping or resizing the image, using a different image file format that may reduce the storage requirements of the image due to the use of data compression, or changing the color depth of the image. In fact, depending on the capability of the image manipulation program you are using, you may be able to combine two or more techniques to significantly reduce image storage requirements and transmission time. As the proof of the pudding is in the eating, let us investigate the effect obtained from altering images. In doing so we will need an image-manipulation program. The program that we will use in this section is WebImage from Group 42 of Milford, Ohio. A trialware copy of this program is included on the CD-ROM accompanying this book, and the installation and operation of this program is described in more detail later in this book.

Image cropping

When we discuss image cropping, we normally associate this term with the direct removal of parts or portions of an image to

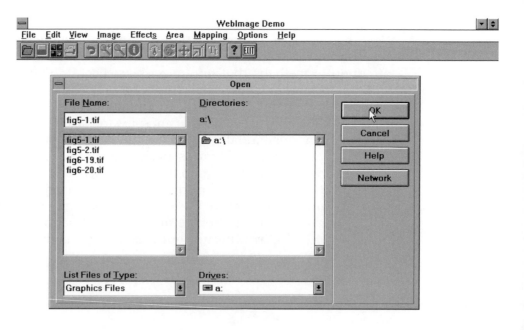

Figure 5.4   Using the WebImage program to retrieve a previously acquired TIF image that was shown in Figure 5.1

reduce its size. Although not actually cropping, the WebImage program provides users with the ability to easily remove a portion of an image and convert it into a new image. As this technique can be considered to represent the effect of image cropping, we will use this program to illustrate how you can adjust an image to an application through cropping. For this first example, we will assume that the image of this author's daughter, as illustrated in Figure 5.1, is to be used with a visual personnel database, photo ID or similar application.

Figure 5.4 illustrates the main window of the Group 42 WebImage program after the Open option was selected from the program's File menu. In the example shown in Figure 5.4, the file fig5-1.TIF was selected to be retrieved from drive a. In effect we are preparing to use the WebImage program to retrieve the image of the author's daughter illustrated in Figure 5.1.

Once you display an image in the program, cropping is initiated by a simple point, click and drag operation. First, you point the

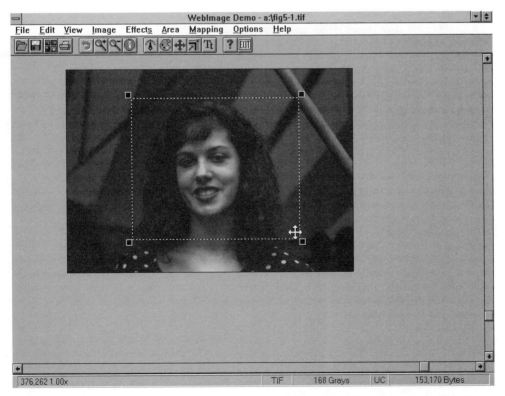

Figure 5.5 Through the use of a point, click and drag operation you can use the Web-Image program to select a portion of an image to form a new image

cursor to the edge of the area that you wish to crop. In actuality, as we will note, the program supports cropping by cutting a portion of an image to the Windows clipboard. Next, you click your mouse button and drag it to form the area that you wish to remove from the image to form a new image. Figure 5.5 illustrates the previously described point, click and drag operation used to form a rectangular area that is more suitable for a photo ID application than the entire image. Once you have selected a portion of an image, the program will paste it to the Windows clipboard. To use that image you would select the New option from the program's File menu. This action results in the display of a dialog box labelled New as illustrated in Figure 5.6. In examining the dialog box that was superimposed on the image, note that the type button for Paste from Clipboard was selected. When you click on the OK button, the program will paste the selected portion of the image previously stored in the Windows clipboard to the screen. Figure 5.7 illustrates the result of the previously described operation.

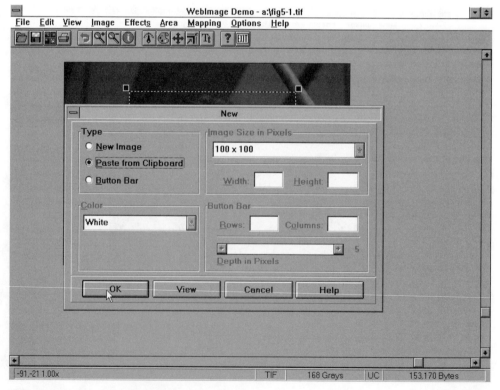

Figure 5.6 Selecting the WebImage program New entry from the File menu enables a previously cut portion of an image to be pasted to the screen

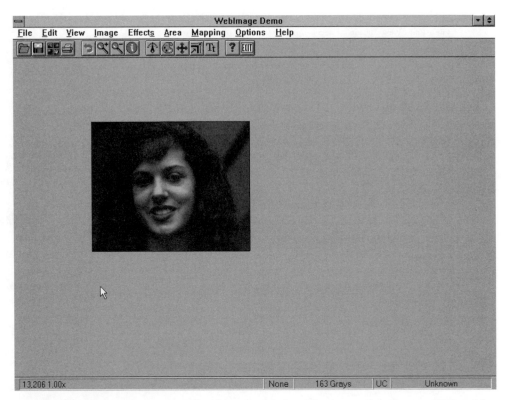

Figure 5.7 The resulting cropped image obtained from pasting the contents of the clipboard to the screen

If you compare the sizes of Figures 5.1 and 5.7, it is obvious that the latter is substantially smaller. Although Figure 5.1 might be more suitable for a magazine, the reduced image shown in Figure 5.7 should be sufficient for a personnel database that requires the inclusion of a picture of an employee, a photo ID application used by the corporate security department, or a similar application. Thus, you could use the SAVE AS option from the program's File menu to save the newly revised and reduced image.

To determine the effect of cropping, we can select the Information entry from the program's View menu. Doing so results in the display of a dialog box appropriately labelled Information which displays a variety of information about the image, including its dimensions, aspect ratio, color depth (which is called Shades by the program) and file size. As indicated in Figure 5.8, the previously described cropping operation reduced the size of the TIF file to 55 254 bytes from its original size of 153 170 bytes. Although this is a significant reduction, note that the file was stored as a TIF file without compression. Thus, a second image

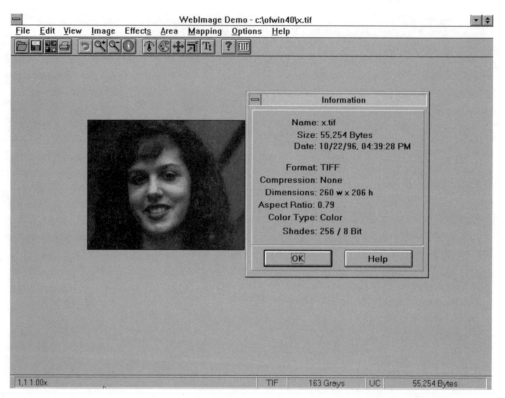

Figure 5.8   Displaying information about the cropped image, including its file size

manipulation technique that warrants discussion, as its effect may further reduce the storage requirements and transmission time to move an image across a LAN, is the use of a file format that supports compression.

Using data compression

When considering the potential use of compression, we must consider two key factors: the capability of the application client program to view different image file formats, and the effect obtained from the use of different types of compression. Concerning the latter, this can include the suitability of lossy and lossless compression, as well as the use of an appropriate quality factor if a lossy compression method is used.

To illustrate the effect obtained from the compression of the previously cropped image, we will again use the WebImage program. As the original program and its cropped reduced image were both stored using a non-compressed TIF file format, it would be logical to explore the effect obtained by storing the image in a compressed TIF file format.

As indicated earlier in this book when we discussed the TIF file format, that format supports a number of compression options, including LZW coding. Unfortunately, the WebImage program limits its support of TIF to either no compression or the use of RLE compression. This is indicated in Figure 5.9, which illustrates the display of the program's Save Options dialog box after the Save option was selected from the program's File menu with TIFF as the type of file selected for listing. The latter is indicated by the presence of TIFF, (*.tif) in the box labelled List Files of Type in the background dialog box. In actuality, although RLE Packbits is displayed in the box labelled Compression in the Save Options dialog box, there is a second option for TIF files supported by the program. That option is None, which when selected results in the file being stored as a non-compressed TIF file.

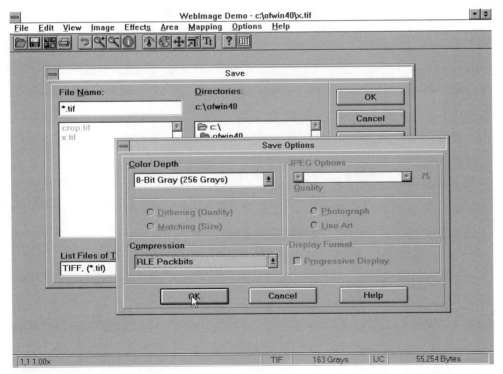

Figure 5.9 Using WebImage to store an image as a compressed TIF file using RLE compression

Figure 5.10 illustrates the effect obtained by storing the cropped, non-compressed TIF file using RLE compression. Note that the file size is 59 024 bytes. In comparison, the non-compressed TIF file of the image shown in Figure 5.8 required 55.254 bytes of storage. In effect, the use of RLE compression resulted in an expansion of the amount of storage required for storing the image. The use of the WebImage program to compress an image vividly illustrates two important concepts. First, to effectively consider a large number of image manipulation options you will probably need to consider the use of two or more image-manipulation programs. Although many programs are very comprehensive, they are usually not all-inclusive. As previously indicated, the WebImage program is limited to supporting two types of TIF file formats, no compression and the use of RLE compression. Thus, if you want to use a TIF file format but wish to use a more effective data compression technique, you would need to consider the use of a different image manipulation program.

To illustrate the use of a second image manipulation program, this author used the commercial Collage Image Manager program

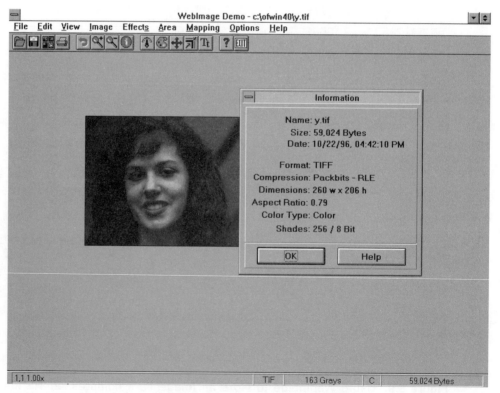

Figure 5.10   Viewing the storage effect obtained by using RLE compression

to open the previously created cropped TIF file produced by the tradeware WebImage program. Once the file had been opened, it was saved using a TIFF image format and LZW compression. Figure 5.11 illustrates the display of the cropped image whose filename was changed to LZW.TIF. It also illustrates the display of a dialog box labelled Image Information which indicates that the use of LZW compression reduced the storage requirements of the previously cropped image to 40 289 bytes.

Although the use of LZW compression had a significantly better effect than the use of RLE compression, we may not wish to stop at this point. After all, we need to consider the application and the file format or formats that the client image viewer supports. Determining answers to those two questions may allow us to proceed further. Assuming that this is not a medical application where we cannot afford the loss of a bit, and also assuming that the client image viewer supports JPEG, we can examine the effect obtained by using a lossy compression method. In doing so, we will return to the use of the WebImage program.

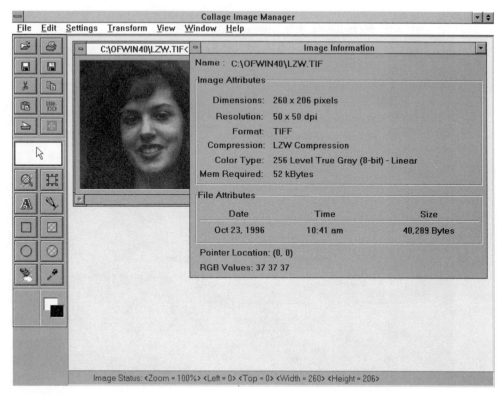

Figure 5.11 Using Collage Image Manager to convert the image into a LZW compressed file format and view the storage requirements of the converted file

Figure 5.12 illustrate the WebImage Save Options dialog box after JPEG was selected as the file type to be saved. In examining the Save Options dialog box, note the sliding bar for Quality which enables a user to directly specify a quality value for use by JPEG compression. In the example shown in Figure 5.12, this author slid the bar to the right to obtain a quality value of 75. Although a lower quality value would result in an additional degree of data reduction, the tradeoff between image clarity and data storage commonly results in the use of a quality value between 50 and 75 being suitable for most applications. When a quality value of 75 was used, the resulting image was reduced to requiring 6852 bytes of storage as illustrated in Figure 5.13. In fact, although the use of a quality value of 50 further reduced the size of the file to 4048 bytes, the background of the image was blending into the foreground hair of the author's daughter. Although still recognizable, the further reduction in the data storage requirements of the file by 2804 bytes from the 6852 bytes required when the image was stored in a JPEG file format using a quality factor of 75, is probably not worth the

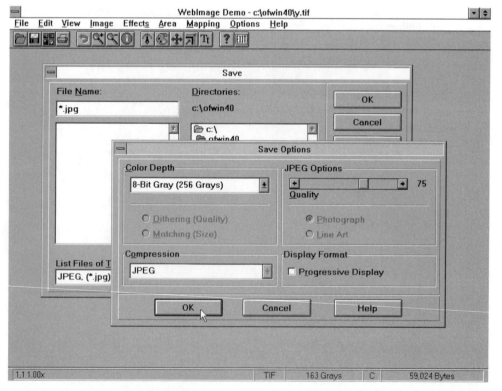

Figure 5.12 Using WebImage to store a file in a JPEG file format using a quality value of 75

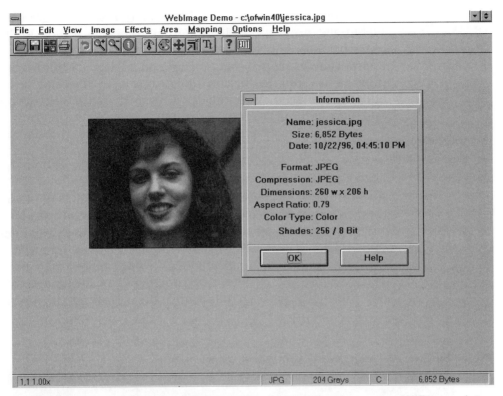

Figure 5.13 Viewing the result obtained from converting the cropped TIF image into a JPEG image using a quality value of 75

minimal additional data reduction. After all, we originally began with an image that required 153 170 bytes of storage and reduced it to 6852 bytes.

Color depth adjustment

When we reviewed basic image concepts earlier in this book, we noted that the storage associated with an image is proportional to its color depth. Often we will acquire an image from a source over which we have no control, such as the purchase of public domain images contained on a CD-ROM. Other times we may directly acquire an image, and only after it has been acquired realize that its color depth exceeds that necessary for the intended application. For either situation it becomes possible to reduce the storage and transmission time associated with an image by changing its color depth.

To illustrate the effect of color reduction we will again use the WebImage program. Selecting the program's Color Reduction entry from its Image menu results in the display of a dialog box appropriately labelled Color Reduction. Figure 5.14 illustrates the program's Color Reduction dialog box with its color reduction options displayed. Although not shown in Figure 5.14, WebImage supports two color reduction methods, dithering and diffusion. Dithering produces a high-quality color reduction through the use of combining colors and patterns to represent different colors. This method is more suitable for operation on images with a high color content, such as photographs. In comparison, diffusion results in the program selecting the closest color value for each pixel to reduce the color depth.

The original image was scanned using 8 bits per pixel to obtain 256 shades of gray. We will alter the image by reducing its color depth. As indicated by the highlighted bar placed over the entry 16 Grays, we will reduce the color depth of the image to 4 bits per pixel. Figure 5.15 illustrates the resulting display of the

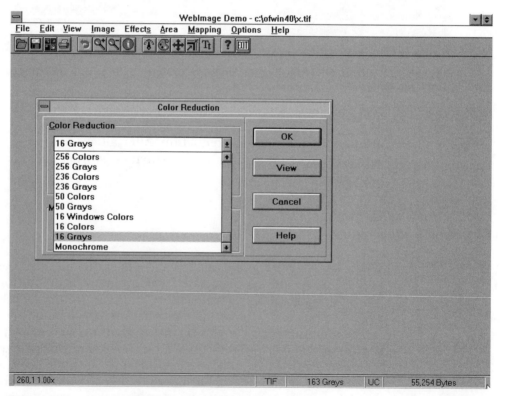

Figure 5.14 Using WebImage to change the color depth of a previously acquired image

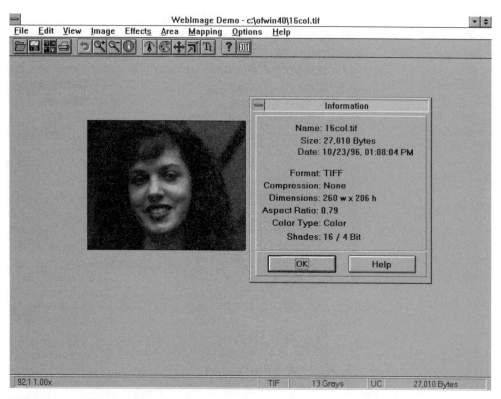

Figure 5.15 Viewing the cropped image after its color depth was reduced to four bits per pixel

reduced color depth cropped image of the author's daughter. Note that the file size of the TIF image was reduced to 27 010 bytes and that the resulting image was stored in a non-compressed TIF format. As application program viewers support different types of file formats, there is no one best format nor is there one recommended color depth. This means that you may have to devote a period of time to performing a trial and error process by converting a sample image into other file formats, with or without adjusting the color depth or image size. To provide readers with a summary of the effect of different image manipulation techniques applied to Figure 5.1, Table 5.2 summarizes those techniques and the resulting file size for each technique. In examining the entries in Table 5.2 readers should note that two additional entries were added that have not previously been discussed. Those entries are the cropped JPEG images with a 4-bit color depth and quality values of 75 and 50. The file size for each image was determined by converting the cropped non-compressed TIF image that had its color depth reduced to 4 bits to JPEG using quality values of 75

Table 5.2 Comparing the manipulation of Figure 5.1

Image	Description	File	Storage
Non-compressed	TIF image	153 170	8-bit color depth
Cropped non-compressed	TIF image	55 254	8-bit color depth
Cropped non-compressed	TIF image	27 010	4-bit color depth
Cropped RLE compressed	TIF image	59 024	8-bit color depth
Cropped LZW compressed	TIF image	40 289	8-bit color depth
Cropped	JPEG image	6 852	8-bit color depth, quality value = 75
Cropped	JPEG image	8 327	4-bit color depth, quality value = 75
Cropped	JPEG image	4 048	8-bit color depth, quality value = 50
Cropped	JPEG image	6 713	4-bit color depth, quality value = 50

and 50, respectively. This additional pair of conversions was performed to illustrate another image storage issue that you should consider. That is, dithering while reducing the color depth and data storage requirements of non-compressed images does not compress as well as a non-dithered image. This becomes apparent when comparing the file storage requirements of the 4- and 8-bit color depth JPEG images listed in Table 5.2. This also means that in general you should consider reducing the color depth of an image when it will be used in an application that does not support a compressed image file format. If your application image viewer can support one or more compressed image file formats, you should convert the image directly into a compressed file format rather than reducing its color depth.

Using thumbnails

In concluding our examination of software-based techniques that can facilitate the use of images on LANs, we turn our attention to the use of thumbnails. A thumbnail can be considered to represent a scaled and reduced representation of a previously created image. Some image manipulation programs provide users with the ability to specify the resolution and size of thumbnails, whereas other programs use a fixed size and resolution. A series of thumbnails used as an entity is referred to as a thumbnail catalog. Through the use of thumbnail catalogs you can provide persons that require the use of images with the ability to preview many images instead of needing to transfer each image onto their computer on an individual basis. This enables persons to avoid downloading images that they may not require, as well as providing them with the ability to focus their efforts on those images of a more immediate interest. Thus, the use of thumbnail catalogs can reduce the transfer of unnecessary image files as well as boosting the productivity of persons that work with a series of images. To illustrate the use of thumbnails and how thumbnail catalogs can provide a more effective use of LAN bandwidth, we will use the thumbnail feature built into WebImage.

To display a sequence of thumbnails, you would select the Select Thumbnails entry from the program's file menu. After selecting an appropriate directory path, the program will display the number of images not found in the database as a prompt to add those images to the thumbnail database. Figure 5.16 illustrates the display of

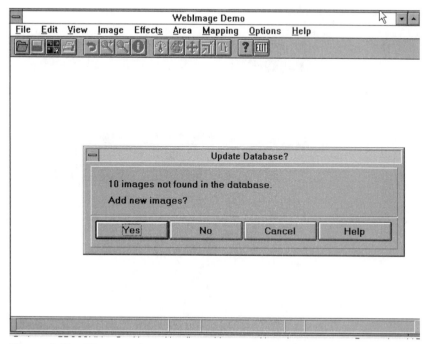

Figure 5.16 The creation of a thumbnail catalog or its update is accomplished through the use of the WebImage Update Database dialog box

the program's Update Database dialog box. Once you click on the button labelled OK, the program will initiate the creation of thumbnails associated with each image in the previously selected path, constructing a catalog of selected images.

Figure 5.17 illustrates the display of one of a sequence of ten dialog boxes that the program displays as it created thumbnails of each of the ten images in the selected directory. Each dialog box is labeled Thumbnail Status and it indicates the name of the file it is using for the creation of the thumbnail and the percentage of the thumbnail creation process that has been completed. In Figure 5.17 the Thumbnail Status dialog box indicates that 69% of the creation of the thumbnail representing Fig5-8.TIF had been accomplished when the screen was captured.

Through the use of the thumbnail creation process applied to a directory containing ten images, a catalog or database of thumbnails was created. This catalog, which is illustrated in Figure 5.18, can be stored on any directory including a network directory. Through the use of thumbnail catalogs, network users can view a large number of images without having to download each image. Then, when a particular image arouses their curiosity for more detailed information, they can double-click on the image to download the full image from a file server.

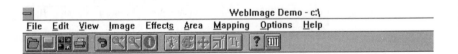

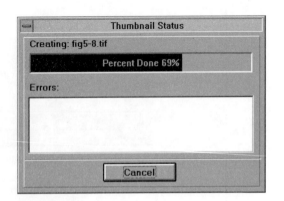

Figure 5.17 The WebImage program uses a Thumbnail Status dialog box to display the status of thumbnails being created for each image

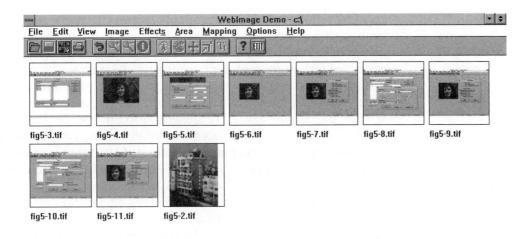

Figure 5.18   Viewing a thumbnail containing thumbnails of ten images

The thumbnail catalog shown in Figure 5.18 required approximately 100 Kbytes of storage. In comparison, the ten full images represented by the thumbnails required over 950 Kbytes of storage. Thus, the ability to display thumbnails as a preview mechanism may reduce the use of network bandwidth, as well as boosting user productivity. Concerning the latter, the thumbnail catalog allows users to see a group of images at one time without needing to individually download each image. This enables users to determine whether or not they wish to view the full image, and it avoids the necessity of time delays associated with the downloading of large images that may not be wanted by the user once he or she has viewed it. Thus, the use of thumbnail catalogues could enhance the effectiveness and efficiency of a real estate visual database as well as other types of related image.

The thumbnail catalog illustrated in Figure 5.18 contained ten images which represents a small fraction of the capability of most programs that support the creation of such catalogs. As most programs have a degree of storage overhead associated with the creation of thumbnail catalogs, this means that as additional thumbnails are created and stored in the catalog, the average amount of storage required per thumbnail decreases. This also means that although a catalog containing 10 images might require 100 Kbytes of storage, a catalog containing 50 thumbnails may only expand its storage requirements to 250 Kbytes of storage. As just one full image could represent 500 Kbytes, or even 1 or 2 Mbytes of storage, a 50 image thumbnail catalog could provide a viable substitute for the expedient viewing of 100 Mbytes of images!

To illustrate how a thumbnail catalog can facilitate the use of a real estate image application, let us assume that all the thumbnail images shown in Figure 5.18 represent various apartment buildings for a commercial real estate database. To view the full image of a particular property the user would select the image of interest. Figure 5.19 illustrates the selection of the photograph of the apartment building in Tel Aviv. By double-clicking on that image, the full image would be downloaded onto the client computer operating the WebImage program.

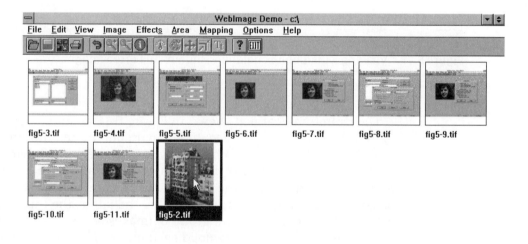

Figure 5.19   Selecting a thumbnail image will result in the display of the full image

By double-clicking on the thumbnail labelled Fig5-2.TIF, the program will automatically retrieve and display the image associated with the thumbnail. Figure 5.20 illustrates the result of the double-clicking operation on the thumbnail labeled Fig5-2.TIF. In this example, WebImage displays only a portion of the image because its original size was 632 by 784 pixels, which exceeds the display's reviewing area.

It should be noted that the ability to use thumbnail catalogs from within an application program will depend on the ability of the application program to support thumbnails. Unfortunately, most image manipulation program developers that include a thumbnail catalog creation capability use a proprietary format to store their catalogs. Thus, it could be difficult, if not impossible, for you to use an application program that has an image thumbnail viewing capability with a thumbnail catalog created by another program. Although this can represent a problem to many potential thumbnail users, the use of a bit of ingenuity may alleviate this problem and allow you to use an image manipulation program directly as a supplement to an existing non-image application program. For

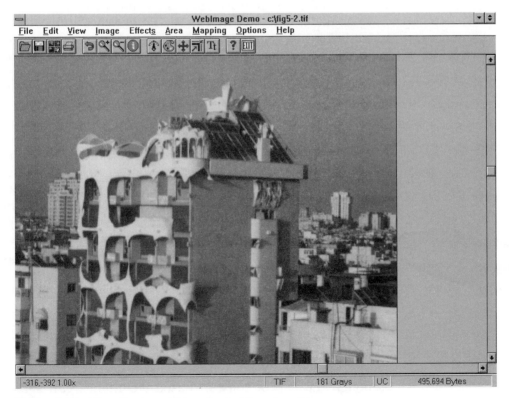

Figure 5.20 The display of the selected thumbnail image

example, returning to real estate, let us assume you are a estate agent who takes a digital camera with you when you visit listings. Assuming that your Multiple Listing Service (MLS) does not support the use of thumbnail images, you could store the photographs that you took with your digital camera on your organization's LAN file server, using an MLS code number as the file name or prefix for each photograph. If you took more than one picture of each property, you could use a coding scheme to associate the type of image with a particular property. For example, if the MLS property number was 456789, you might consider the suffix EF for exterior front, EB for exterior back, IK for interior kitchen and so on, to explain the view associated with each image. Then you could store the images on your network file server and use WebImage or a similar program that supports thumbnail catalogs as well as one that is network-aware to create thumbnail catalogs that could be used as a property preview mechanism. Concerning network awareness, WebImage is fully network-aware, allowing you to use the program with images stored on network servers as well as to save modified images to network-based devices. This networking capability is obtained by selecting a button labelled Network that is included in Open, Save, and Save As dialog boxes as well as the Build New Catalog dialog box displayed when you use the program to create a thumbnail catalog. Concerning the latter, Figure 5.21 illustrates the use of the program's Network Drive Connection dialog box after network drive F was selected. By simply closing the dialog box and clicking on the OK button contained on the Build New Catalog box in the background of Figure 5.21, the program would construct a thumbnail catalog of images in the selected network path. You can also use the Network Drive connection dialog box to map a network directory path to a drive letter, delete a previously established network drive mapping, and perform other network-related actions without having to exit the program. Thus, the use of a network aware image manipulation program that supports the creation of thumbnail catalogs may be sufficient by itself for many image-based applications. Now that we have an appreciation for software-based techniques that can be used to enhance the use of LAN-based images, let us turn our attention to hardware-based techniques.

Hardware-based techniques

Although the use of software-based techniques can serve as a mechanism to minimize the effect of images on LAN perfor-

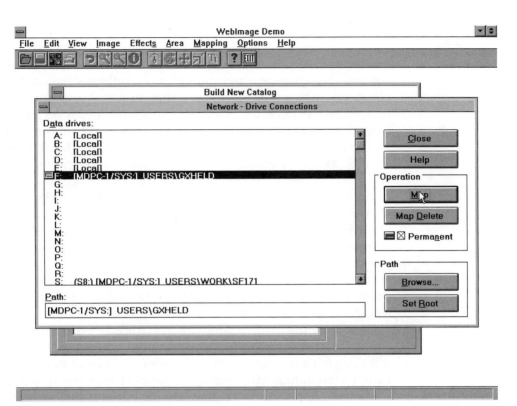

Figure 5.21 Through the Network Drive Connections dialog box you can select a previously mapped network drive or establish a new network mapping

mance, there are certain network environments where you will need to turn to more expensive hardware-based techniques. Those environments include a saturation of existing network bandwidth due to the general level of network utilization, or an anticipated or actual growth in the use of images that exceeds the capability of software-based techniques to minimize their effect to a reasonable level of degraded LAN performance. For such situations you will more than likely need to consider the use of one or more hardware-based techniques. Those techniques can include network segmentation, reducing the collision window on Ethernet LANs, using a high-speed backbone to interconnect segmented networks, using intelligent switching hubs to provide multiple simultaneous connections to overcome shared media single connection constraints, and the replacement of one type of network by another higher-capacity LAN. In the remainder of this section we will examine each of those hardware-based techniques.

Network segmentation

There are certain network environments where the use of software-based image manipulation techniques has a minimal effect on network performance due to the overall transfer of images on the network. For such situations, it may be possible to consider the segmentation of the network if the requirement for image access is limited to a defined subset of office employees. To illustrate this concept, let us assume that a real estate office has 25 employees, of which 12 are real estate agents, while the remaining employees are secretaries, advertising clerks, contract document preparers, managers and other types of supporting staff. Suppose that the real estate office has a common local area network which agents use to preview images of homes stored on a server whenever a potential customer is brought into the office to preview listings before going with the agent to actually view properties. Let us also assume that other office employees access wordprocessing spreadsheets and database programs.

If a few agents bring customers into the office and begin to 'flip' through stored property images, this activity can seriously interfere with interactive query–response communications associated with filling out forms, a common real estate document processing application. To alleviate the effect of the transmission of images on client–server interactive query–response applications you can consider segmenting the network. For the real estate office, the top of Figure 5.22 illustrates a common Ethernet network before segmentation, and the lower portion of that illustration indicates the possible segmentation of the network. In this example, it was assumed that the network used Novell's NetWare operating system.

One of the lesser known features of NetWare is the ability of a file server to perform internal bridging to interconnect two local area network segments. Although Novell refers to this as LAN Routing, in effect the file server functions as a transparent bridge to connect the two network segments together. The only hardware required to perform bridging is a second Ethernet adapter card which can be obtained for less than $100. Thus the cost to implement segmentation is most nominal.

In examining the segmented LAN illustrated at the bottom of Figure 5.22, note that the 12 agents that require the downloading of images from the server were placed on a common segment. As an agent typically converses with the client as images are being painted on their monitor, a delay of a few seconds or more may hardly be noticeable nor seriously impair the productivity of the real estate agent as one agent's image requests compete with

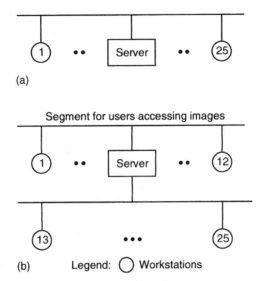

Figure 5.22 Segmented Ethernet LAN to minimize the effect of image transmissions

those of other agents. In comparison, suppose several interactive query–responses to create documents on the server are in progress when one or a few agents initiate image transfers. Although the Ethernet CSMA/CD access protocol, including its exponential random backoff period when a collision occurs, acts as an arbitrator to provide each station with equivalent LAN access, the large number of frames required to transport each image can only interfere with the frames carrying interactive query–response traffic used to prepare a document. When this occurs, a workstation user will encounter random delays as he or she enters various information fields on the document. Not only will typing be awkward but, in addition, the random delays will adversely affect the effort of the typist entering data into the document.

By segmenting the Ethernet LAN shown in the top portion of Figure 5.22 into two segments, the effect of transmitting images upon interactive text query–response applications is minimized. Thus, network segmentation represents a viable method that you should consider to minimize the effect of image transmission on an existing local area network.

Reducing the collision window on Ethernet LANs

In a CSMA/CD network stations first 'listen' to the network prior to transmitting. If two stations are located relatively far from one another the probability that one station fails to hear and then initiates its own transmission increases. Thus, as the distance between stations increases, the probability of the occurrence of collisions increases. As the occurrence of a collision results in the first station to detect the collision transmitting a jam signal that precludes transmissions from other workstations for the duration of the Ethernet frame, LAN performance decreases as collisions increase. Thus, there is a high correlation between reduced Ethernet network performance and the distance between very active workstations.

Although it is probably impractical to relocate more than a few very active workstations to reduce the distance between stations, it may be relatively easy to reduce the cable distance between the two most active stations on most networks. Those two stations are commonly the file server and print server, which are normally the most active devices on a local area network.

The top of Figure 5.23 illustrates an Ethernet network where a file server is located at a relatively long cable distance from a print server, resulting in a long collision window. The lower portion of

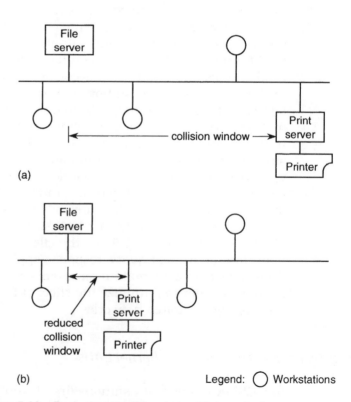

Figure 5.23 Reducing the Ethernet collision window

Figure 5.23 illustrates a reduction in the collision window between highly active network stations obtained by simply recabling one server closer to the other. Based on the use of performance monitoring equipment by this author, a reduction in network utilization between 5 and 10% has been obtained by relocating a few active workstations closer to one another.

Using a fibre backbone

When constructing a local internetwork consisting of several linked LANs within a building, one method of minimizing the effect of image traffic on other network applications is to place image applications on image servers located on a separate high-speed network.

Figure 5.24 illustrates the use of a fibre distributed data interface (FDDI) backbone ring consisting of two image servers that can be accessed from workstations located on several Ethernet and Token Ring networks through local bridges linking those networks to the FDDI ring. By using the FDDI ring for image applications, the 100 Mbps operating rate of FDDI provides a delivery mechanism that allows workstation users on multiple

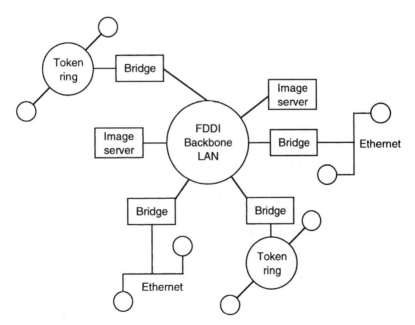

Fig. 5.24 Using a high-speed FDDI backbone

lower-operating-rate LANs to simultaneously access image applications without experienceing network delays.

For example, one network user on each LAN illustrated in Figure 5.24 accesses the same image application on an image server connected to the FDDI backbone LAN. If each Token Ring network operates at 16 Mbps, and each Ethernet operates at 10 Mbps the composite transfer rate from the FDDI network to each of the lower-operating-rate LANs bridged to that network is 52 Mbps. Because the FDDI network operates at 100 Mbps, it can simultaneously present images to network users on each of the four LANs without any internetwork bottlenecks occurring.

Another advantage associated with using an FDDI backbone restricted to supporting image servers and bridges is economics. This configuration minimizes the requirement for using more expensive FDDI adapter cards to one card per image server and one card per bridge. In comparison, upgrading an existing network to FDDI would require replacing each workstation's existing network adapter card with a more expensive FDDI adapter card.

To illustrate the potential cost savings, assume that each Ethernet and Token Ring network has 100 workstations, resulting in a total of 400 adapter cards, including two image servers that would require replacement if each existing LAN was replaced by a common FDDI network. Because FDDI adapter cards cost approximately $800, this replacement would result in the expenditure of $320 000. In comparison, the acquisition of four bridges and six FDDI adapter cards would cost less than $20 000.

Using intelligent switching hubs

The use of intelligent switching hubs permits multiple connections between network clients and network servers which boosts network bandwidth beyond that obtainable from the use of a shared media network. To illustrate how the use of an intelligent switching hub can reduce the effect of the transfer of bandwidth-intensive images on network users, consider Figure 5.25 which shows the use of an eight-port switching hub. In this example, the eight-port hub is used as a mechanism to interconnect two individual LAN users, four Ethernet segments containing groups of stations, and two network servers. Note that, although each LAN user and network segment operates at 10 Mbps, the connections between the switch and each server are 100BASE-T Fast Ethernet connections that operate at 100 Mbps. In addition, note

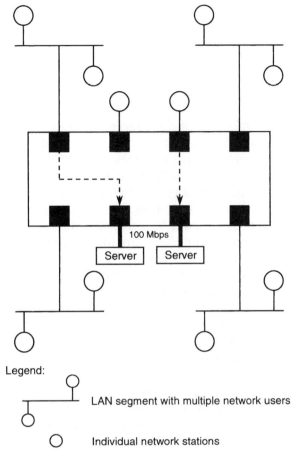

Figure 5.25 Using a switching hub. Through the use of a switching hub multiple simultaneous client–server communications becomes possible, multiplying the amount of available bandwidth

that two simultaneous communications sessions are shown in progress in Figure 5.25. In comparison, if the two servers and each network segment and individual LAN users were connected to a common network, only one session at a time would be possible. Thus, the use of a switching hub to connect multiple servers to either individual stations or multiple stations on multiple segments, supports multiple communications sessions that multiplies the potential bandwidth of the network. In the example illustrated in Figure 5.25, the use of Fast Ethernet connections between the hub and each server further enhances potential network throughput. This is because the use of Fast Ethernet enables servers to respond more quickly to client queries. This in turn permits each server to become available sooner to service another query which enhances transmission through a switching hub-based network.

Infrastructure change

In concluding this chapter, we will discuss one additional hardware-based technique that can be used to enhance network performance to accommodate the transmission of images. This technique, which we have saved for last, is usually the most expensive and time-consuming method that you can employ to increase LAN bandwidth. If it appears that none of the other techniques will provide a sufficient level of network performance, you should then consider changing your network infrastructure.

A complete change to an existing LAN infrastructure can be very expensive and time-consuming. This is because it usually involves replacing existing network cabling, adapter cards and hubs. Examples of LAN infrastructure changes can include migrating from 16 Mbps Token Ring to 100 Mbps FDDI, a change from 10 Mbps Ethernet to 100 Mbps Fast Ethernet, or replacing either a Token Ring or Ethernet network by 25 or 155 Mbps ATM-based LANs. Although any of those infrastructure changes may be able to acommodate your LAN-based imaging requirements, they should be considered as a last resort and implemented only if software and other hardware-based techniques prove insufficient.

6

WEB-BASED IMAGES

Today it is difficult to pick up a magazine or newspaper, or to turn on the television, and fail to note one or more references to the World Wide Web. This book is no exception because the importance of the Web is such that any book covering network-based images must discuss this topic. In this chapter we will turn our attention to the use of images in World Wide Web documents known as Web pages. In doing so we will first focus our attention on obtaining an overview of the composition of HyperText Markup Language (HTML) documents. Once we have an appreciation for the creation and use of fundamental HTML statements, we will then turn our attention to the use of specific HTML image-related statements where we will note the effect of different statements on the storage and transmission time of images. Although the focus of this chapter is on the use of images on Web servers, readers should note that Chapter 7 continues our discussion of Web servers. In that chapter we will examine the effect of images when determining an appropriate method to connect a server to the Internet.

6.1 HTML OVERVIEW

The HyperText Markup Language (HTML) provides a standardized mechanism for creating a wide range of documents that are displayed by Web browsers. Such documents can range in scope from Web pages that simply display text and graphics to pages that contain menus and include linked groups of information which, when selected, can result in database queries, initiation of e-mail and other functions.

An HTML document is similar to a text file and can be created using any text editor or wordprocessor so long as the use of the latter results in the generation of an ASCII file without word-processor control codes. Although an HTML document is created as ASCII text, it includes tags that define how information is viewed, as well as operations that are performed when certain types of displayed information are clicked on. Although you can use a text editor or wordprocessor to create HTML documents, many persons prefer to use HTML editors that simplify the document-creation process. In this chapter we will primarily use a text editor to generate HTML examples; however, it should be noted that there are a large number of HTML editors that you can use, including one built into Netscape Gold. Once an HTML document has been created, its subsequent viewing requires the use of a browser. As Netscape Gold includes both a browser and an HTML editor, its use facilitates the Web page development process as you can use that program to create, view, modify and review HTML documents.

HTML structure

An HTML document consists of a hierarchy of elements known as markup tags. Each tag has a name and may have one or more attributes and/or contents. Figure 6.1 illustrates an example of

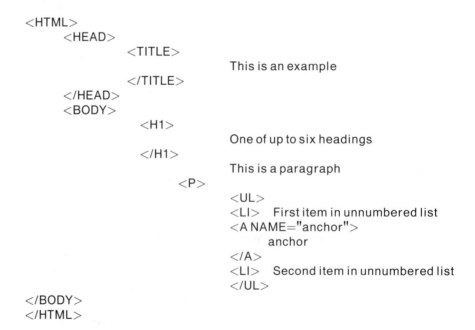

```
<HTML>
    <HEAD>
            <TITLE>
                        This is an example
            </TITLE>
    </HEAD>
    <BODY>
            <H1>
                        One of up to six headings
            </H1>
                        This is a paragraph
                <P>
                        <UL>
                        <LI>   First item in unnumbered list
                        <A NAME="anchor">
                            anchor
                        </A>
                        <LI>   Second item in unnumbered list
                        </UL>
    </BODY>
    </HTML>
```

Figure 6.1   HTML instance example

the generic structure of an HTML document which represents a sequence of markup tags. In examining the HTML document structure shown in Figure 6.1, note that the indentations were established to illustrate the relationship of document tags. When creating an HTML document, indentations are irrelevant. In fact, you do not have to place markup tags on separate lines and can string them together or alternate their placement. For example, you can end the text of a paragraph by the use of the markup tag <P> instead of placing it on a separate line. Now that we have a basic understanding of the structure of an HTML document, let us focus our attention on the use of some common markup tags and the information that can be displayed by a tag or sequence of tags.

Markup tags

A markup tag can be considered to represent an identifier that informs a viewer how to display text. HTML tags consist of a left angular bracket < followed by a keyword or mnemonic known as a directive and closed by a right angular bracket >.

As indicated from an examination of the entries in Figure 6.1, most markup tags are paired, with the terminator markup tag identical to the initiator tag with the exception of a prefix forward slash / which precedes the text or mnemonic enclosed in the brackets. The forward slash tells the viewer that the prior viewing operation established by a preceding markup tag is terminated. The primary exception to the pairing of markup tags under an early version of HTML was the <P> end-of-paragraph tag, because its use eliminates the need for a </P> tag. In fact, there was no such markup tag as </P> under early versions of HTML.

Text and mnemonic entries in markup tags are case-insensitive. Thus, <TITLE>, <title> and <Title> are completely equivalent to one another.

The primary method used to view a HTML document is through the use of a browser. Prior to 1994 the most popular browser was Mosaic, developed at the National Super Computer Center located in Urbana, IL. Many of the persons responsible for the development of Mosaic left the National Super Computer Center to form Netscape Communications Corporation which currently dominates the Web browser market; however, the Internet Explorer from Microsoft has gained market acceptance, and both support a core set of HTML statements that make HTML programming for viewing by both popular browsers possible. Each of those software programs, as well as other browsers, were developed with coding which acts on HTML markup tags. Because the HTML standard is in a state of evolution, not all tags are supported by all browsers. If a browser does not support a specific markup tag it should ignore the tag. However, some versions of browsers that ignore unknown markup tags also omit the display of information within pairs of unknown tag entries that can result in gaps in the display of information. Fortunately, most browsers with a release date after mid-1995 should support all standardized HTML markup tags specified in HTML 2.0. Thus, a check of browsers' release dates should provide you with the ability to note whether or not the browser will support the complete HTML 2.0 markup tag standard which is supported by most Web page development tools. Although HTML versions 3.0 through 3.2 were being implemented when this book was prepared, the majority of Web documents conform to HTML 2.0. Now that we have an appreciation for the release dates of browsers, let us focus our attention on the function and utilization of some of the key markup tags.

The directive for the title tag is <TITLE>. Normally the title is positioned on the first line of a document and the text of the title is used to identify the contents or use of the document. On encountering the TITLE tag, a browser displays the information contained by the tag pair <TITLE> and </TITLE> on the top line of the screen.

Figure 6.2 illustrates two examples of the use of the HTML title tag. Note that the first example included the title terminator tag on the same line, and the second example used separate lines for each markup tag and the text of the title to be displayed. Both methods are acceptable, although the second method may be preferable due to its clarity when contained in an extensive document.

Six levels of heading, numbered 1 through 6, are supported by HTML. Heading 1 is the most prominent, and succeeding numbers represent less prominent headings.

When encountered by a browser, the text within beginning and ending heading markup tags is displayed in larger and/or bolder

```
<TITLE>Personnel Database Query</TITLE>

or          <TITLE>
            Personal Database Query
            </TITLE>
```

Text can be positioned with beginning and ending tags on one line or on separate lines.

Figure 6.2   Title tag examples

fonts than the normal body text. The first heading in a document should be tagged <H1>, although you can use any heading level if you so desire.

In constructing an HTML document you can consider using a level 1 heading as being equivalent to a chapter, and level 2 headings could be used to identify major sections within the chapter. Then levels 3–6 could be used to visually identify different areas within a section or to highlight specific information that you wish users to visually note.

The format of the heading tag is indicated below:

<Hn> Text of heading </Hn>

where n is a number between 1 and 6 and it represents the level of the heading. For example, to display the string Retrieve data by Social Security Number as a level 2 heading you would enter the following HTML statement:

<H2> Retrieve data by Social Security Number </H2>

A paragraph in HTML is similar to a paragraph in a book or article, consisting of sentences of related information. When creating a paragraph it is important to note that carriage returns and white spaces in HTML files are not significant. In addition, word wrapping can occur at any point in the source file that you create.

When creating a paragraph you should separate each paragraph from a succeeding paragraph by the <P> tag. Otherwise, the viewer will treat separately entered paragraphs as one large paragraph which is not your intention but may result in a visually unappealing display. The following example illustrates the use of title, heading and paragraph tags within an HTML document:

<TITLE> Personnel Data Base Query </TITLE>
<H1> Retrieve data by Social Security Number </H1>
<P1> To retrieve personnel data based upon the social security number of an employee click on the number icon. </P>

In general you can omit the use of the </P> closing tag because when a browser encounters a subsequent <P> tag, that tag implies that there is an end to the previous paragraph. Through the use of the ALIGN = alignment attribute within the <P> tag, you can control the location where a paragraph is displayed. For example,

<P ALIGN = CENTER>
This paragraph is centered.
</P>

would result in the text 'This paragraph is centered.' being centered on this display.

The key to the versatility of hypertext documents is the ability to create documents that enable a user to jump to a different section within the document or to specific sections in other documents. To accomplish this task required the development of a standardized mechanism for addressing documents. That mechanism is known as the Uniform Resource Locator (URL). Thus, before examining the use of anchors within a HTML document let us briefly discuss URLs.

The format of a URL is indicated below:

scheme://host.domain[:port]/path/filename

In this format the scheme can have a variety of values. For example, the scheme can be coded as 'file' to denote a file on a local computer or on an anonymous ftp server, 'http' (hypertext transmission protocol) to access a file on a World Wide Web server, 'gopher' to access a file on a Gopher server, or 'WAIS' to access a file on a WAIS server. Table 6.1 lists the scheme entries currently supported by URLs. The port number is normally not required; however, it should be used if the manager of the destination address changes port addresses from their default values as a mechanism to obtain an additional degree of security.

As an example of the use of a URL, assume that you wish to access the file personnel.jan on the FTP server whose address is opm.macon.gov. If the file is located in the directory PERSONNEL you would use the following URL:

ftp://opm.macon.gov/PERSONNEL/personnel.jan

Table 6.1 Uniform resource locator schemes

Scheme entry	Scheme access
ftp	File transfer protocol
gopher	Gopher protocol
mailto	Electronic mail address
mid	Message identifiers for electronic mail
cid	Content identifiers for MIME body part
news	Usenet news
nntp	Usenet news for local NNTP access
prospero	Access using prospero protocol
telnet	Interactive telnet session
tn3270	Interactive telnet 3270 session
WAIS	Wide area information servers
http	Hypertext transfer protocol

Now that we have an appreciation for the construction of URLs, let us return to the main topic of this section, anchors.

As previously mentioned, anchors can be used to establish links to a section in the same or a different document. To illustrate the creation and use of anchors, let us assume that you wish to establish a link from document 1 to a specific section in document 2. To do so you must first establish a named anchor in document 2 and then create the link in document 1 which references the previously named anchor. For example, assume that you want to add an anchor named SSN to document 2. Then you would insert text similar to the following example. Note that you can insert text before and after the beginning and ending anchor brackets.

< A NAME="SSN" > Retrieval based upon Social Security Number < /A >

To reference the location in document 2 you must include its filename as well as the named anchor in document 1. In doing so, you would separate the URL for document 2 by a hash mark #. Thus, one possible entry in document 1 to establish a link to document 2 would be as follows:

To access the data based upon the employee, click on SSN.
< A HREF="URL#SSN" > SSN < /a >

In the prior example, URL would be replaced by the actual Uniform Resource Locator to denote the location of document 2. Then clicking on the mnemonic SSN in document 1 would send the reader directly to the words 'Retrieval based upon Social Security Number' in document 2.

When establishing anchors within the same document you would follow the same method previously described; however, you could replace the URL by the mnemonic html as you are simply performing a hypertext jump within the same document.

HTML supports three types of lists: unnumbered, numbered and definition. Unnumbered lists represent unordered information, and numbered lists display ordered information. The third type of list, the definition list, is used to display alternative descriptive titles and descriptive descriptions. Each type of list is created using an appropriate set of markup tags. For example, an unnumbered list commences with an opening list <ul> tag. Next, each individual item in the list is prefixed with the list identifier <li> tag; however, no closing list identifier tag is required. Instead, the unnumbered list is completed with a closing list </ul> tag. The following example illustrates the creation of a four item unnumbered list:

```
<UL>
<LI> 1960–1969
<LI> 1970–1979
<LI> 1980–1989
<LI> 1990–1999
</UL>
```

When a browser displays the entries in an unnumbered list, it prefixes each entry with a bullet. Thus, the output would be visually displayed as follows:

. 1960–1969
. 1970–1979
. 1980–1989
. 1990–1999

A numbered list, which is also referred to as an ordered list, is identical to an unnumbered list but uses the tag pair <OL> and </OL> in place of <UL> and </UL>. Thus, the items in a numbered list are tagged using the same <LI> tag. Returning to our previous decade listings example, we would code a numbered list as follows:

```
<OL> 1960–1969
<LI> 1970–1979
<LI> 1980–1989
<LI> 1990–1999
</OL>
```

The preceding numbered list would be displayed as follows:

1. 1960–1969
2. 1970–1979
3. 1980–1989
4. 1990–1999

The definition list uses the tag pair <DL> and </DL> as a prefix and a suffix to the list. Within that tag pair the tags <DT> and <DD> are used alternatively to denote the definition term and the definition. Shortening our previously used decade list into a definition list, we might consider the following coded example of a definition list:

```
<DL>
<DT>  1960–1969
<DD>  The decade of the hippies
<DT>  1970–1979
<DD>  Ten years of growth
</DL>
```

The use of a browser to view the definition list would result in the following display:

```
1960–1969
   The decade of the hippies
1970–1979
   Ten years of growth
```

HTML supports a number of different text formats. Its pre-formatted <PRE> tag is used to identify a block of text that should be displayed with spaces, lines and tabs functioning in their normal manner. Other text formats supported by HTML include block quotations and character formatting. The use of block quotes provides you with the ability to display quotations in a separate block on the screen. Character formatting provides you with the ability to display individual words or sentences in italics, bold, fixed width font or as the result of a specially defined tag. By using block quotes and character formatting you can customize the display to ensure that it is both visually appealing and easy to use. Text will be displayed in italics by using the <I> and </I> tags. Similarly, the use of the <B> and </B> tags results in the display of text in bold. One special tag that deserves note is the <BR> tag that generates a line break without white space occurring between lines. For example:

```
Gilbert Held<BR>
Macon, GA 31210<BR>
```

would generate this author's name followed by his city, state, and zip code on the next line without space between lines.

In concluding this overview of HTML let us discuss the use of in-line images and briefly defer a detailed description until the next section. Thus, we will defer until the next section how images can turn an ordinary document into a visually appealing display. In addition, the old adage 'a picture is worth a thousand words' becomes fully appreciated when you note that through HTML you can perform such functions as displaying the photograph of employees while displaying database information and similar integrated image and text displays.

Embedded images can be inserted into a document display through the use of an IMG tag containing different attributes such as SRC and ALIGN. The SRC (source) attribute is used to denote the URL of the document to be embedded. ALIGN takes values of TOP, MIDDLE or BOTTOM and defines the location where graphics and text should be aligned vertically. To illustrate the use of graphics within a display, assume that you wish to display the contents of the GIF file named SSCARD.GIF which contains an image of a Social Security Card. To do so you could enter the following HTML statement where the text outside of the brackets simply displays information around the image:

An Employee's Social Security Card
< IMG SRC="SSCARD.GIF" > must be examined at time of employment.

Now that we have a basic understanding of HTML, let us turn our attention to examining in detail those image-related statements that provide the visual impact associated with visiting different Web sites.

6.2 INLINE IMAGES

The term inline image refers to the placement of images next to text, and it is used as a catch-all phrase associated with the display of images on Web pages. In this section we will examine in some detail the use of the IMG tag and the different attributes associated with that tag under different versions of HTML.

Basic IMG tag format

The basic format of the IMG tag is indicated below:

< IMG SRC="location.ext"[attributes] >

In the preceding format, the location refers the physical path to the file containing the image to be displayed including its file name. The ext entry denotes the extension of the image file that informs a browser how the image stored on the file should be interpreted. Together location.ext represents the URL of the image file.

Image formats

On the World Wide Web GIF and JPG image formats are primarily used, and they are supported by most Web browsers. A third format supported by a few browsers is the X-Bitmap black and white image format which uses the filename extension XBM. As described earlier in this book, GIF uses the LZW lossless compression method and JPG employs a lossy compression method, with both formats considerably reducing the file size and its transmission time. In comparison, XBM represents a non-compressed image file format that is an inefficient method for storing and transmitting images. Owing to this, you should consider converting X-Bitmap images into a different format. In addition, if you can control the use of browser plug-in modules such as when constructing a corporate intranet, you can then consider storing images on your Web server in a more efficient image file format than GIF, such as the Portable Network Graphic (PNG) format. Then you can install a PNG plug-in viewer into the browsers used by your organization to obtain the ability to use intranet browsers to view inline images stored using the PNG format.

Image tag attributes

To effectively examine the use of HTML image tag attributes requires one or more images to work with. Rather than return to the use of the author's dog who is tired by this time, two new images will be used. The first image was taken by this author on a trip provided by the courtesy of the U.S. Army. While on a weekend pass to climb Mt. Ranier in the state of Washington, this author passed an establishment bearing his name. Using his trusty camera, Figure 6.3 illustrates the resulting scanned image of the photograph. Scanned using a 50 by 50 dpi resolution, this 976 by 608 pixel image required 378 392 bytes of storage when stored as a GIF 256 gray scale image. Although this author will place text on a few Web pages that will contain different manipulated images of Figure 6.3, such examples are for illustrative purposes only and have no relationship to the items sold by the pictured establishment.

The second image that we will use in this chapter was taken by this author after he climbed Mt. Ranier. When viewed from a different route on his way back to base, this author wondered why

Figure 6.3 A building bearing the author's name that will be used to illustrate the operation of HTML image tag attributes

he did it! Figure 6.4 illustrates the resulting 256 level gray scan of the photograph whose dimensions were 952 by 536 pixels using a resolution of 50 by 50 dpi. The resulting file required 406 572 bytes of storage when stored as a GIF image.

Figure 6.4 The author's scanned photograph of Mt. Ranier which will be used to illustrate the use of HTML image tag attributes

The ALIGN attribute

Under HTML 2.0 three alignment options are supported for use with the ALIGN attribute. Those options are ALIGN = top, ALIGN = bottom, and ALIGN = middle, with ALIGN = bottom being the default.

The use of the ALIGN = top option results in text aligned with the top of the image being displayed. However, after one line is displayed aligned with the top of the image, continuing text is displayed at the bottom of the image. Depending on the size of the image, the use of ALIGN = top can result in a considerable amount of white space being viewed.

By default the bottom of an image is aligned with text following the IMG tag statement. However, you can also use the option ALIGN = bottom for consistency when coding HTML statements. The third ALIGN option supported by HTML 2.0 is ALIGN = middle, which centers one line of text on the image, with continuing text displayed below the image. Figure 6.5 lists the contents of a short HTML file developed to generate a Web page containing three images with text aligned at the top, middle, and bottom of succeeding images. Figure 6.6 illustrates the resulting Web page. In examining the HTML listing in Figure 6.5 you will note the use of two additional image tag attributes, WIDTH and HEIGHT, so let us discuss their use.

```
<HTML>
<HEAD>
<TITLE>Gilbert's Garage</TITLE>
</HEAD>
<BODY>
<H2>GILBERT'S GARAGE – YOUR HIGH-TECH STORAGE</H2>
<IMG SRC="GILBERT.GIF" ALIGN=TOP WIDTH=140 HEIGHT=70>
<A <H3>VISIT OUR MODERN FACILITIES</H3>
<H3>We specialize in high technology
to include Web server platforms</H3> </A>
<IMG SRC="GILBERT.GIF" ALIGN=MIDDLE WIDTH=140 HEIGHT=70>
<A <H3>VISIT OUR MODERN FACILITIES</H3>
<H3>We specialize in high technology
to include Web server platforms</H3> </A>
<IMG SRC="GILBERT.GIF" ALIGN=BOTTOM WIDTH=140 HEIGHT=70>
<A <H3>VISIT OUR MODERN FACILITIES</H3>
<H3>We specialize in high technology
to include Web server platforms</H3> </A>
</BODY>
</HTML>
```

Figure 6.5 The contents of a short HTML file that illustrates the use of the IMG tag ALIGN attribute options

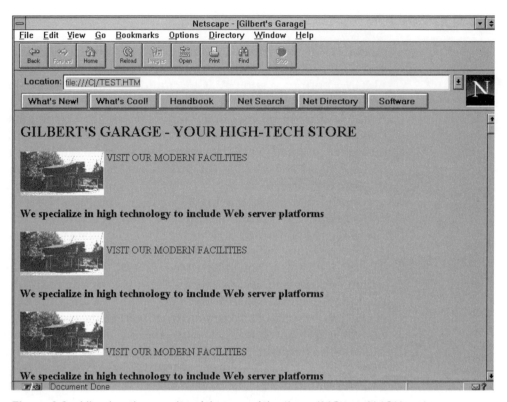

Figure 6.6   Viewing the results of the use of the three IMG tag ALIGN options

WIDTH and HEIGHT attributes

When a browser encounters an image it does not know how much space to leave on a page for the image until it is actually retrieved. Thus, a browser will stop generating text and wait for enough information to be retrieved from the header in the image file. This wait is necessary so that it can determine how much space is needed for the image before continuing to display text as it downloads and displays the image. Depending on the size of the image and its color depth as well as your network connection operating rate, this can result in a significant delay in the display of a Web page. Recognizing this problem, Netscape Navigator Version 1.1 included support for image tag HEIGHT and WIDTH attributes which are now supported by most browsers as well as incorporated in the HTML 3.0 specification. Through the use of the HEIGHT and WIDTH attributes you can scale an image so that it will be displayed at a different size than the actual image. Both attributes are used with the equals sign (=) followed by the number of pixels you want for the image width and height on a Web page.

Through the use of many browsers as well as image editing programs, you can determine the size of an image. Concerning the former, some browsers will indicate the size of an image in the title bar if you load it directly into a window, alleviating the necessity to use an editing program. For either situation it is important to note that the resolution of the commonly used VGA display is 640 by 480 pixels which means that you probably want to specify a subset of that width and height. As indicated in Figure 6.5, the width and height were set to 140 and 70, respectively, to enable three images to be displayed on the same page for illustrative purposes. Also note that you can use several IMG tag attributes together so long as you include a space between attributes. This is indicated by the following statement that was one of three IMG tag statements included in Figure 6.5:

<IMG SCR="GILBERT.GIF" ALIGN=TOP WIDTH=140 HEIGHT=70>

A word of caution is in order when using WIDTH and HEIGHT attributes within an image tag. When you specify values that differ from the actual width and/or height of the image, the browser will use the specified values in the IMG tag and scale the image to fit. This can result in a considerable amount of image distortion when you specify values considerably different from the aspect ratio of the stored image.

Although the use of TOP, BOTTOM and MIDDLE values with the ALIGN attribute provides some basic vertical alignment options, those options do not allow text to 'float' around an image. Recognizing this limitation, many browsers now support the use of ALIGN attribute values of LEFT and RIGHT.

LEFT and RIGHT attribute values

The use of the LEFT ALIGN attribute value places an image on the left of a Web page and allows following text to flow around the image to its right. Similarly, the use of RIGHT results in the image floating to the right on a Web page, with any text following the image or included in a paragraph containing the image flowing to the left of the image.

Figure 6.7 illustrates the construction of a short segment of HTML code, including the use of the IMG tag with an ALIGN= RIGHT attribute value using WIDTH and HEIGHT pixel values of 300 and 200, respectively. Figure 6.8 illustrates the resulting display appearing on the author's Netscape browser. In examining the HTML code listed in Figure 6.7, note that the <BR> tag

```
<HTML>
<HEAD>
<TITLE>Gilbert's Garage</TITLE>
</HEAD>
<BODY>
<H1>GILBERT'S GARAGE – YOUR HIGH-TECH STORAGE</H1>
<IMG SRC="GILBERT.GIF" ALIGN=RIGHT WIDTH=300 HEIGHT=200>
<H1>VISIT OUR MODERN FACILITIES</H1>
<H2>We specialize in high technology<BR>
to include Web server platforms</H2>
</BODY>
</HTML>
```

Figure 6.7 HTML code developed to illustrate the use of the ALIGN=
RIGHT image tag attribute value

forces a line break similar to a hard carriage return. By simply
changing ALIGN=RIGHT to ALIGN=LEFT we can reverse the
positions of the image and text. The result of this change is
illustrated in Figure 6.9.

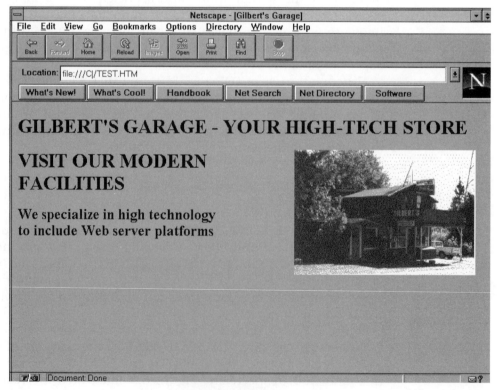

Figure 6.8 Using the ALIGN=RIGHT image tag attribute value to let text flow to the
left of an image

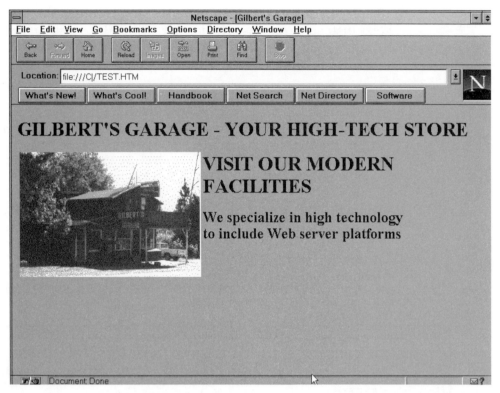

Figure 6.9 Through the use of the ALIGN=LEFT image tag attribute, value text will flow to the right of an image

Before continuing our discussion of the use of the IMG tag, a slight digression concerning browser incompatibilities with respect to image viewing is warranted. Although Netscape Navigator currently has between 70 and 80% of the browser market, it is important to note that only a few years ago Mosaic was the most commonly used browser. In addition, some browsers cannot display images, while those that do include the ability of the user to turn off image loading to facilitate expediting the viewing of text on Web pages, an important consideration when using a low-speed modem to access a Web server. Thus, a discussion of the use of the ALT attribute which enables text associated with an image to be displayed when the image cannot be displayed is warranted.

The ALT attribute

Through the use of the ALT attribute you can associate text with an image which will be displayed by browsers incapable of displaying images, or by browsers that are configured to display text instead of images. The format of the ALT attribute is:

ALT="[text]"

To illustrate the use of the ALT attribute, let us assume that you wish to associate the text alternative Gilbert's Garage with the image GILBERT.GIF. To do so, you would specify the ALT attribute within the IMG tag as follows:

< IMG SRC="GILBERT.GIF" ALT="[Gilbert's Garage]" >

Through the use of the preceding ALT attribute, the string Gilbert's Garage would be displayed within a pair of brackets when viewed by a browser that does not support graphics or when viewed by a graphics capable browser whose graphics display option is disabled.

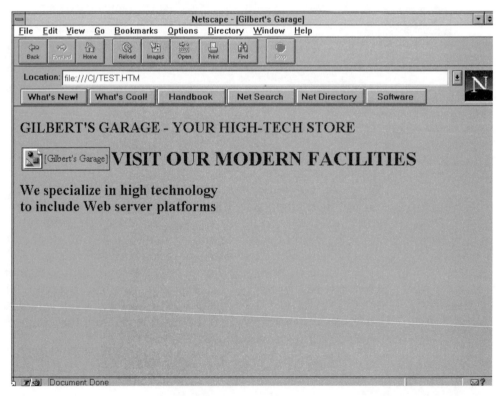

Figure 6.10 If you turn off the Auto Load Images option from Netscape's Options menu, Netscape displays a small icon to the left of the text generated by the IMG tag ALT attribute. Clicking on that icon results in the loading and display of the image

Different browsers use ATL in different ways; however, the basic use of the attribute remains the same. For example, consider Netscape's Navigator browser. If you use the browser's Options menu and click on the Auto Load Images option to disable that option, the browser displays a small icon to the left of the text contained in brackets to indicate that an image is associated with the text, as illustrated in Figure 6.10. That image can then be loaded and viewed by clicking on the icon.

You can include multiple attributes within an IMG tag, including the ALT attribute. Thus, the use of ALIGN, WIDTH and HEIGHT attributes can be included with the use of the ALT attribute. For example, the following IMG tag could be used:

```
< IMG SRC="GILBERT.GIF" ALIGN=LEFT WIDTH=300
HEIGHT=150 ALT="[Gilbert's Garage]" >
```

If you fail to use the ALT attribute, the string [IMAGE] will be displayed. On occasion, if you are using decorative images that have no underlying meaning you may wish to consider a small trick to hide the image notation from appearing on the screen display of a non-graphical browser whose graphics display is disabled. To do so, you would use the ALT attribute to assign a null or empty description to the picture. For example, returning to the author's garage image, we could encode the IMG tag as follows:

```
< IMG SRC="GILBERT.GIF" ALT=" " >
```

Spacing

If you examine the display shown in Figure 6.9 you will note that text is positioned almost directly on the image floated to the left of the display. To prevent text from being directly positioned on floated images, Netscape introduced two attributes to the IMG tag, HSPACE and VSPACE. HSPACE, a mnemonic for horizontal spacing, is used to define the amount of space in pixels that are left between an image and surrounding text. Similarly, VSPACE is used to define the number of pixels to be left at the top and bottom of an image. Both HSPACE and VSPACE accept numeric parameters and the format of each attribute is shown below, with n replaced by the number of pixels of space to be left horizontally or vertically around an image:

```
HSPACE = n
VSPACE = n
```

Returning to the author's garage, to leave 20 pixels space around all sides of the image we would use the following IMG tag:

```
< IMG SRC="GILBERT.GIF" ALT="[Gilbert's Garage]"
WIDTH=300 HEIGHT=150 HSPACE=20 VSPACE=20
```

Although you can put HSPACE and VSPACE to good use to separate text from graphics, you do not need to experiment with their use to center an image. To display an image without any associated text in the center of a Web page you can use the ALIGN=CENTER attribute associated with the paragraph tag. Again returning to the author's garage image, you could use the following HTML code to center the image:

```
< P ALIGN=CENTER >
< IMG SRC="GILBERT.GIF" HEIGHT=300 WIDTH=150 >
< /P >
```

The preceding code would center the image whose height is 300 pixels and width is 150 pixels on the display, with subsequent text placed below the image left justified.

Border attribute

One often overlooked method that can be used to highlight images is obtained through the use of the BORDER attribute. This attribute has the format:

```
BORDER=value
```

where the value represents the thickness of the border that the browser draws around an image. As the browser generates the border the use of the BORDER attribute has no effect on the time required to download an image regardless of the size of the image or thickness of the border. Figure 6.11 illustrates the results obtained from the use of HEIGHT, WIDTH, CENTER, and BORDER attributes. The actual HTML statements used to generate the centered image surrounded with a large border were as follows:

```
< P ALIGN = CENTER >
< IMG SRC="Gilbert.Gif" border=10 width=300 height=200 >
< /P >
```

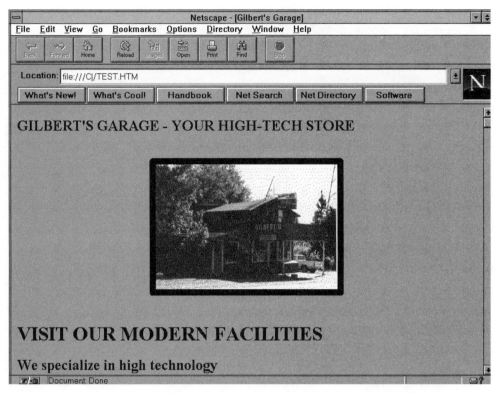

Figure 6.11   Using the BORDER and CENTER attributes to emphasize an image

Storage size considerations

One of the problems associated with the use of images is their storage requirements, that can considerably differ from their display requirements. Although you can use the HEIGHT and WIDTH attributes to appropriately scale an image to a portion of a Web page, you may have stored an oversized image whose retrieval could result in a degree of ill-will expressed towards you by persons who use a slow modem to connect to your server. To illustrate this problem, consider the display of the author's favorite garage aligned on the left of a Web page and played using a width of 300 pixels and height of 200 pixels. As the image was stored using a resolution of 976 by 608 pixels, let us find out what happens when you click on the right mouse button to obtain the ability to view the image. As previously noted in Chapter 4, pressing the right mouse button when the cursor is placed on an image results in the display of a menu of options associated with the image. Figure 6.12 illustrates the display of the pop-up menu resulting from clicking the right mouse button when the cursor

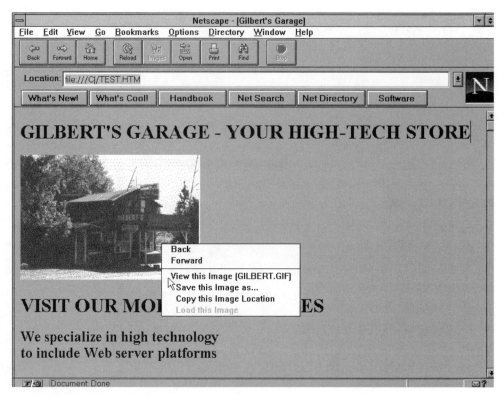

Figure 6.12 Using the right mouse button to display image related options

was placed on the image. Figure 6.13 illustrates the initial display of the image after the View this Image option was selected from the pop-up menu. Note that instead of downloading a 300 by 200 pixel image, the browser downloads the original 976 by 608 pixel image. As previously noted, the GIF image required 378 392 bytes of storage. If the Web browser user were connected to your Web server via a 28.8 Kbps modem connection, he would need to wait $378\,392 \times 8/28\,800$, or 105 seconds, for the image to be downloaded. Owing to this you may wish to consider the use of an image manipulation program to adjust images stored on your Web server.

Using low-resolution images

To facilitate the display of images, Netscape added support for an IMG attribute appropriately using the mnemonic LOWSRC. Through the use of this attribute, Netscape will load and display

Figure 6.13 Viewing an oversized image 'hiding' in the background

a smaller or lower-resolution image, providing the Web page viewer with a general indication of the actual image to be displayed. As the reader is viewing the test and low-resolution image, the real version of the image specified by the SRC attribute will be loaded.

The format of the LOWSRC attribute is LOWSRC="URL", where the URL identifies the location of the image. As an example of its use, consider the following IMG tag, which results in the low-resolution JPG image being displayed before its replacement by the higher resolution GIF image.

< IMG SRC="HIGRES.GIF" LOWSRC="LOWRES.JPG" >

To provide a summary of IMG tag attributes, the seven attributes previously described are listed in Table 6.2. That table also indicates the possible values that can be assigned to each attribute, and it provides a description of the use of the attribute with the indicated value.

Table 6.2 IMG tag attributes.

Attribute	Utilization	Description
ALIGN	ALIGN=top	Aligns current line of text with top of image.
	ALIGN=middle	Aligns current line of text with middle of image.
	ALIGN=bottom	Aligns current line of text with bottom of image.
	ALIGN=left	Floats image to left, text wraps around right side of image.
	ALIGN=right	Floats image to right, text wraps around left side of image.
HSPACE	HSPACE=n	Controls horizontal space in pixels (n) around left and right image.
VSPACE	VSPACE=n	Controls vertical space in pixels (n) above and below image.
HEIGHT	HEIGHT=n	Specifies the height in pixels (n) of the image.
WIDTH	WIDTH=n	Specifies the width in pixels (n) of the image.
BORDER	BORDER=n	Specifies the thickness of a border displayed around an image.
LOWSRC	LOWSRC="URL"	Loads a low-resolution image before the higher-resolution image.

Image links

Although the display of graphics including images are an important part of a Web page, it is the links among and within the page that provides the true 'webbing' within the World Wide Web. This linkage, which is commonly referred to as hypertext, is obtained through the use of anchor tags in a document. Earlier in this chapter we indicated how an anchor could be used to link a word or string to a document that can be in the same directory on one Web server or in a different directory on a different server. In fact, through the use of an anchor you can link a word or string to a file on an FTP server, a connection to a Telnet-based service, a file on a Gopher server and other URL connections. In addition to establishing links using text, you can use anchor tags to associate images with text. In doing so you can provide browser users with the ability to decide if they really want to view a particular image. For example, consider the following HTML statement:

< A HREF="http://www.xyz.com/devel/mountain.gif" >
Click here for Picture (800kb download) </A>

The preceding statement would highlight the text 'Click here for Picture (800 kb download)' which also informs the potential viewer of the image of its size. If a browser user were connected to the Web server via a slow speed modem, they might have second thoughts about clicking on the hyperlink.

In addition to associating images with hyperlinks, you can also use images as hypertext anchors. To do so you would place the IMG tag within the anchor tags. In doing so it is important to understand the placement of images within the anchor tag pair. Thus, let us examine how this is accomplished via the use of generic pictures before creating an example. Using generic terms we can denote the use of a picture within a picture hypertext reference as follows:

```
<A HREF="hiddenpic.ext"> <IMG SRC="displaypic.ext"
ALT="[text]"> </A>
```

Based on the preceding HTML statement the image displaypic will be displayed by the browser. Clicking on that image will then result in the display of the image hiddenpic.ext. Of course, hiddenpic and displaypic.ext would be replaced by the URL of each image. It should be noted that a graphical browser will indicate that an image is a hypertext anchor by boxing the image with a colored or highlighted box.

To illustrate the use of an image as a hypertext anchor, let us display a scaled version of Mt. Ranier but use the image as a hypertext anchor which, when clicked on, generates a full display of the mountain. To do so we could use the following HTML statement:

```
<A HREF="MTRANIER.GIF"> <IMG SRC"MTRANIER.GIF"
WIDTH=150 HEIGHT=75> </A>
```

As many persons may not realize that a boxed image represents a link, you would probably prefer to prefix or suffix the link with a text statement such as:

Click on image to display complete photograph.

Figure 6.14 lists a short segment of HTML code which illustrates the creation of a bordered and centered image whose dimensions are 300 by 200. To provide space around the image, both HSPACE an VSPACE attributes were included within the image tag. Note that the message 'Click on the mountain to view the splendor of a 500 Kbyte image' both instructs the person viewing the page how

```
<HTML>
<HEAD>
<TITLE>Outdoor Cafe</TITLE>
<H2>View nature </H2>
<p align=center>
<A href="mtranier.gif">
<IMG SRC="mtranier.gif" border=10 width=300 height=200 ALT="[Mt.
Ranier]" hspace=20 vspace=20>
</a>
</p>
</BODY>
</HTML>
```

Figure 6.14 HTML coding to display the bordered and centered image shown in Figure 6.15. When the image is clicked on, the full image is downloaded and displayed

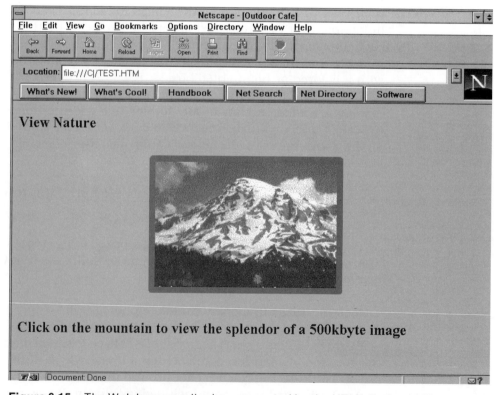

Figure 6.15 The Web browser display generated by the HTML listing in Figure 6.14. When the image is clicked on, the full image linked to this sized image will be downloaded and displayed

to display the full image and informs them of the size of the file containing the image. Figure 6.15 illustrates the Web browser display generated by the HTML listing in Figure 6.14.

Background and interlaced images

Prior to examining the actual protocol used to transport Web pages and methods that can be used to enhance their transmission efficiency, two additional types of images warrant review. Those types of image are background and interlaced images.

A background image represents an image that is loaded and used as a background for a Web page. To use an image as a background is accomplished through the use of the BACKGROUND tag as an addition to the <BODY> tag that surrounds your HTML statements. Figure 6.16 lists a short section of HTML code that generates a Web page using the author's photograph of Mt. Ranier as the page background with subsequent text displayed in the foreground. Figure 6.17 illustrates the display of the resulting Web page.

In examining Figure 6.17, which illustrates the use of the author's photograph of Mt. Ranier as a Web page background, you will note that the color of the background to a degree obstructs the message. Unfortunately, many Web page designers fail to realize that the background display should be in the background and not distract from the rest of the display. In fact, the use of improper backgrounds probably represents one of the most abused Web design techniques, especially when you consider the time required to load background images.

As we have just discussed Web page backgrounds, another term warrants an elaboration. That term is tiling. If the selected image is smaller than the background area of a Web page, a

```
<HTML>
<HEAD>
<TITLE>Gilbert's Camping Supplies</TITLE>
<BODY BACKGROUND="MTRANIER.GIF">
<H2>GILBERT'S CAMPING SUPPLIES – YOUR ONESTOP SHOP</H2>
<H3>We specialize in boots, thermals and gloves<BR>
to keep you warm as you climb</H3>
</BODY>
</HTML>
```

Figure 6.16 HTML coding to display Mt. Ranier as a Web page background

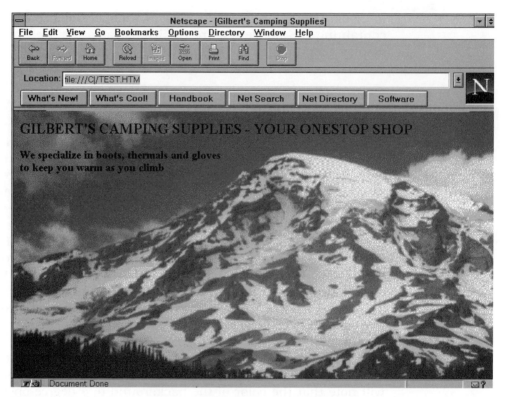

Figure 6.17   Mt. Ranier displayed as a Web page background

browser will use the downloaded background image to fill the page by repeating it across and down to fill the window. Thus, from an operational perspective, if you intend to use a background image you should consider using a relatively small image that the browser will repeat across the Web page. Otherwise users accessing your Web server via low-speed modem connections may need to wait a considerable amount of time for a large background image to be downloaded and displayed.

A second type of graphic image that warrants attention is the interlaced image. When an interlaced image is used, it is 'painted' on the screen by a server transmitting the image in a format by which alternating lines of the image are displayed. One example of a commonly used interlaced image format is GIF89a. When you store an image as an interlaced sub-format using the GIF89a file format, its subsequent display results in every fourth line of the image, starting at the top, being transmitted. When one-quarter of the image is received it is displayed, providing the browser user with a rough indication of the image. The second quarter of the image is then painted on the screen followed by the third and fourth portions. Through the use of an interlaced image a browser user can observe the image while it is being loaded rather than waiting as the image is slowly displayed on a line by line basis. Although the use of interlaced images does not alter the amount of time required to fully display an image, it can enhance the productivity of users viewing your Web site. This is because, instead of waiting for the full display of an image as it is decoded and displayed incrementally, the viewer can make a decision earlier as to whether or not the image is of interest and whether they should continue to view it or do something else.

You can create interlaced images through the use of an appropriate image manipulation program. One such program is PaintShop Pro, a popular shareware program from JASC Inc. of Eden Prairie, MN. If you are using this program you would first open a previously acquired image. Then you would use the Save as option using GIF as the file type and version 89a-Interlaced as the file sub-format. Figure 6.18 illustrates the use of the program's Save as option to save the author's picture of Mt. Ranier as an

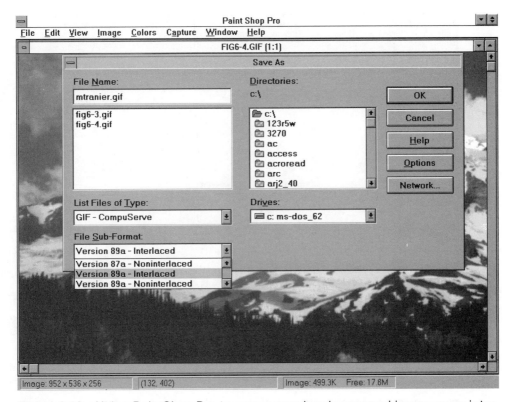

Figure 6.18 Using PaintShop Pro to save a previously scanned image as an interlaced GIF89a image

interlaced image. Figures 6.19 and 6.20 illustrate the progressive display of the interlaced image while it was being painted on the author's browser.

If you focus your attention on the lower left corner of the browser's edge that contains the display of the interlaced image, you will note that in Figure 6.19 the browser indicates that 0% of 399 K has been displayed. Although not entirely accurate because the first series of scan lines for most of the image has been painted on the screen, you can note from the blocks that constitute the image that it is very rough. In Figure 6.20 the screen was captured a few moments later, with the first sequence of lines from the first decoding state fully displayed in the browser's viewing area. At this point the browser indicates that 5% of the 399 Kbyte file has been received. As the display of subsequent scan lines occurs, the image blocks shown in Figures 6.19 and 6.20 begin to obtain a better visual resolution and the image of Mt. Ranier becomes much more recognizable.

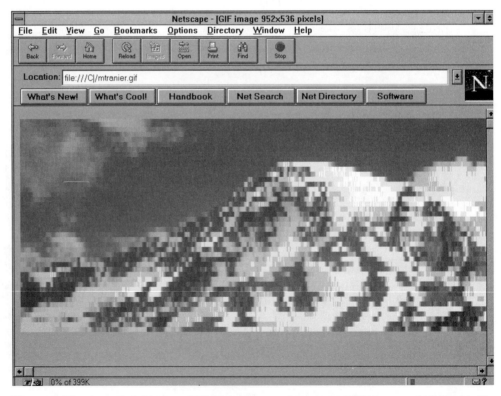

Figure 6.19 A partial display of Mt. Ranier as an interlaced GIF image. At this point in time only a portion of the first scan line out of every four lines has been displayed

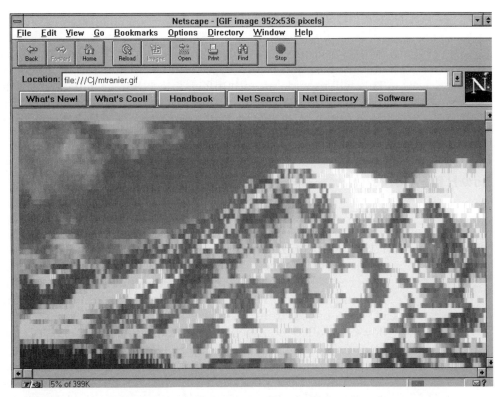

Figure 6.20 Another partial display of Mt. Ranier as an interlaced GIf image. At this point in time the first scan line out of every four liens has been fully displayed and a small number of second scan lines were displayed. The latter can be observed by comparing the top portion of the clouds in Figure 6.19 to Figure 6.20

Although the display of an interlaced image originally resulted in a Venetian blind type of display as alternate quarters are received and displayed, several browser developers changed their code to replicate the available lines of the interlaced image during the first three of four decoding stages. This results in a browser user initially viewing a complete, but not properly displayed, image whose quality is enhanced as additional quarters of the image are painted, until the entire image is displayed at its full resolution.

Although interlaced GIF is commonly used on Web documents, it is not the only type of interlaced image. Progressive JPEG, a recent variant of JPEG, results in the display of an image at progressively higher detail similar to an interlaced GIF. Although support for progressive JPEG was limited to a few browsers when this book was prepared, you can reasonably expect its use to expand. This is due to the fact that the storage and transmission time required for JPEG images are considerably less than for equivalent but not identical GIF images.

6.3 HTTP OPERATION

In this section we will first obtain an appreciation of the client–server operation of the HyperText Transmission Protocol (HTTP). This will provide us with information that we will use to obtain an understanding of how we can enhance the efficiency of the Web page design process, as well as how to use a technique known as client pull to enhance the appeal of Web pages.

How HTTP works

HTTP represents the application layer protocol that is transported via the Transmission Control Protocol/Internet Protocol (TCP/IP). HTTP is a simple request/response protocol that operates as a client–server process, with browsers functioning as the client, whereas a Web server program functions as the server that responds to client requests. TCP/IP provides the reliable transfer of Web pages from the server to the browser via a window flow-controlled connection-oriented protocol. Once TCP/IP has established a connection, data transfer initiallly commences rather slowly because the window size that governs the number of unacknowledged packets that can be outstanding is slowly increased up to its maximum. If a browser user is located some distance from the server, or if the transmission facility has a large amount of network traffic, the acknowledgement of the previously received packets will encounter a delay in reaching the server, momentarily resulting in a pause in the client–server communications session. Further exacerbating this condition is the fact that an HTTP connection is limited to a single request for information. This means that a Web page that contains 10 graphics requires a series of 11 connections, with each connection resulting in a sequence of small delays. As the client-server connection is initiated, the client initiates a request and the server performs a disk search to locate and retrieve the requested information, transmits the requested information, and closes the connection. Figure 6.21 illustrates the sequence of HTTP client–server packet exchanges to retrieve a document containing several inline images.

Based on the preceding information you can considerably minimize the time required to retrieve a Web page by reducing the number of graphics on a page. Although you may prefer to retain several small images due to their display of important information rather than minimize Web page loading time, with the appropriate image manipulation program you may be able to

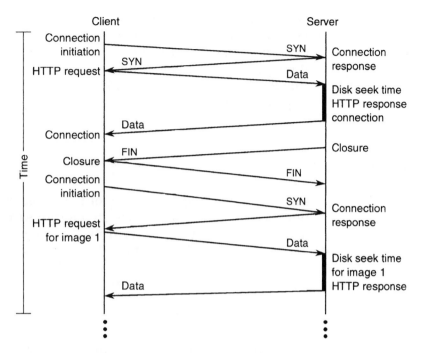

Figure 6.21   HTTP client–server packet exchange sequence

literally have your cake and eat it, too. To do so, you can consider retrieving several small images, positioning them together in an appropriate manner, and then storing the collage as a single image. Doing so would result in the browser needing to use a single request instead of multiple requests to retrieve one image instead of a sequence of images. Now that we have an understanding of how HTTP operates, let us conclude this section by examining a technique that delays HTTP sessions and whose use can result in more dynamically appealing Web pages. This technique was pioneered by Netscape and is referred to as client-pull.

Client-pull

Client-pull represents a mechanism first introduced with Netscape Navigation version 1.1 that enables a programmer to sequence the display of a series of URLs without requiring the intervention of the browser user. Client-pull obtains its name as the browser client pulls the same or a new URL from a host after a specified period of time. Through the use of client-pull you can periodically update the display of stock market information, network usage statistics, or other dynamically changing information. In addition, you can even 'walk' an image across a screen through its use, although the storage of multiple images within GIF89a files is more suitable for generating an animation sequence of images.

The META tag

Client-pull is based on the use of the HTML META tag. When a browser encounters a META tag it interprets the tag's parameters as an HTTP header generated by the server and performs the operation specified in the tag. The format of META tag is shown below. Note that you must use the META tag right after the HEAD tag to generate a client-pull operation.

```
< META HTTP-EQUIV="Refresh" CONTENT="n;URL" >
```

In the preceding format n defines the time the browser waits before reloading the same or a different Web page. If a URL is not specified, the browser reloads the same Web page. To illustrate the use of the META tag, let us examine a few examples of its use. For example, if you want a browser to reload a Web page every five seconds you would use the following HTML code.

```
< HTML >
< HEAD >
< META HTTP-EQUIV="Refresh" CONTENT=5 >
< TITLE > Client-pull demo < /TITLE >
< /HEAD >
```

Now suppose that we specify a different document within the META tag, modifying the tag as indicated below:

```
< META HTTP-EQUIV="Refresh" CONTENT="1";
http://www.xyz.com/devel/wow.htm" >
```

Changing the META tag would cause the browser to load the document wow.htm after a delay of one second, initiating a loop in which the browser loads the first page, then the second page and then the first page. The browser will continue this loop until the browser user decides to view a different page or closes their browser.

If you consider using a META tag to perform a client-pull operation, you should probably consider minimizing or eliminating the use of images on the pulled page. This is because the delays associated with the transfer of graphics could adversely effect the intended goal of a client-pull operation, such as the near real time display of stock market quotations. Although images can considerably enhance the visibility of a Web page, as with candy too much or improper use can create an undesired effect.

7

SELECTING A
WEB SERVER
CONNECTION RATE

In this chapter we turn our attention to a common problem that organizations can expect to encounter when using the Internet or when constructing a corporate intranet. That problem is the selection of an appropriate transmission rate to connect the organization's Web server to the Internet or to a corporate intranet. As there are differences in the selection process based on the type of network that the Web server will be connected to, we will first examine the process by which we can determine an appropriate connection rate when we wish to connect the server to the Internet. Using the information as a base will enable us to turn our attention to intranet connection considerations in the second part of this chapter.

7.1 INTERNET CONNECTIVITY

The selection of an appropriate Web server connection rate to the Internet involves overcoming two related problems. If the operating rate of the Internet connection is too slow, members of the Internet community accessing your organization's server may become frustrated and terminate their access of information from your Web server site. At the opposite extreme, if your organization's Internet access connection operating rate exceeds the bandwidth required to support an acceptable level of access, you may be wasting corporate funds for an unnecessary level of transmission capacity. Thus, a methodology that can provide you with the ability to select an appropriate Web server connection rate to the Internet can represent a valuable tool to balance user access requirements against the cost associated with providing a connection to the Internet.

Overview

An appreciation for the methods by which a Web server can be connected to the Internet, as well as knowledge about some of the transmission constraints associated with a Web server connection, is facilitated by examining a schematic diagram that represents a typical connection method. Figure 7.1 illustrates the method by which Web servers are normally connected to the Internet. As indicated in Figure 7.1, a Web server will reside on a local area network, with the LAN connected via a router to an Internet access provider. The Internet access provider has a direct connection to a backbone network node on the Internet, commonly using a full T3 or Switched Megabit Data Service (SMDS) connection to provide Internet access for a large group of organizations that obtain Internet access through its connection facilities.

Although an Ethernet bus-based LAN is shown in Figure 7.1, in actuality any type of local area network that can be connected to a router and for which TCP/IP drivers are available can be used by

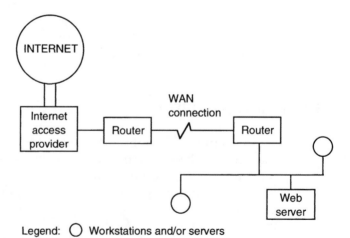

Legend: ○ Workstations and/or servers

Figure 7.1   Web server connection to the Internet. Determining an appropriate operating rate for the WAN connection to the Internet represents a common problem facing organizations that wish to obtain a presence on the World Wide Web

the Web server. Thus, other common LANs used by Web servers include Token-Ring and FDDI as well as the numerous flavors of Ethernet, such as 10BASE-T, 100BASE-T and VG-AnyLAN.

The actual Wide Area Network (WAN) connection between the Internet access provider and the customer can range in scope from low-speed analog leased lines to a variety of digital leased lines. Owing to most Internet access providers recommending a minimum operating rate of 56 Kbps for the WAN connection, only a few access providers offer analog leased line connection options. When offered, the actual operating rate of the WAN connection is commonly limited to 19.2 Kbps or 24.4 Kbps based on bandwidth constraints of a voice grade analog leased line limiting modem operating rates. Concerning digital leased line operating rates, most Internet access providers offer 56 Kbps, fractional T1 in increments of 56 or 64 Kbps to 784 Kbps, full T1, fractional T3 and full T3 connectivity. Although the WAN operating rate serves as a constraint on the ability of users to access information from your organization's Web server, another less recognized but equally important constraint exists that you should consider. That constraint is the traffic on the local area network on which the Web server resides. Thus, although the focus of this chapter is on determining an appropriate WAN operating rate to connect a Web server to the Internet, we will also examine the constraints associated with LAN traffic on the ability of the server to respond to information requests received from the Internet.

WAN connectivity factors

Three key factors will govern the selection of an appropriate operating rate to connect a Web server to the Internet via a wide area network transmission facility. Those factors include the composition of the Web pages residing on a server, the types of pages retrieved by a person accessing the Web server, and the number of hits expected to occur during the busy hour.

A typical Web page consists of a mixture of graphics and text. For example, a university might include a picture of Old Main on the home page in the form of a GIF file consisting of 75 000 bytes of storage supplemented by 500 characters of text that welcomes Internet surfers to the university home page. Thus, this university home page would contain 75 500 bytes that must be transmitted each time a member of the Internet community accesses the home page of the university. By computing the data storage requirements of each page stored on your Web server and by estimating the access distribution of each page, you can compute

Assumptions:

Data Storage	Bytes	Access Percentage
Home Page	120,500	40
Tier 1 Page	80,500	30
Tier 2 Page	60,500	30

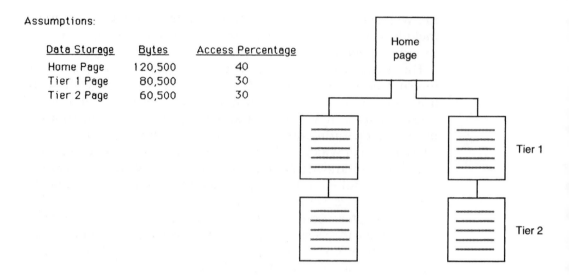

Figure 7.2   Web page relationship

the average number of bytes transmitted in response to each Internet access to your organization's Web server. For example, assume that your organization plans to develop a Web server that will store four distinct Web pages as well as a home page, providing Internet users with the ability to access two types of data from the home page based on the construction of a dual two-tier page relationship under the home page, as illustrated in Figure 7.2. Although the Web home page is always accessed first, from the home page users typically access other server pages based on the hypertext links coded on the home page. Similarly, on accessing different server pages, the ability of a user to jump to other pages on the server is constrained by the links programmed on each page. Thus, the data transmitted in response to each page that an Internet user accesses, as well as the sequence of pages accessed, will probably differ from organization to organization. This means that the example that will be used to compute an appropriate WAN operating rate is for illustrative purposes only.

Performing the required computations

Let us assume that through conversations with other organizations that have implemented Web servers or through information provided by your Internet access provider, you have determined that when Web pages are arranged in a tier structure, access to a home page at the top of the tier represents 40% of all accesses, and the remaining 60% is subdivided by remaining tiers. Let us further assume that your organization's Web page structure will be constructed in two tiers below the home page, with the data storage associated with each page to include text and graphics, as well as the access percentage of each page as indicated in the lower portion of Figure 7.2.

Once you have determined the data storage required for each Web page and the distribution of Internet access by page, you can compute the average number of bytes that will be transmitted from the Web server in response to each hit on your organization's server. Here the term hit represents an access request to a Web page on your server via the use of the HyperText Transmission Protocol (HTTP) using a Uniform Resource Locator (URL) that represents a file stored on your server, which equates to the contents of a Web page.

By using the information contained in the table at the lower portion of Figure 7.2, you can compute the average amount of data that will be transmitted resulting from a hit on the organization's server as follows:

$$\text{Average data transmitted} = \sum_{e=I}^{3} \text{bytes on page I}$$

$$* \text{ access percentage for page I}$$

Substitution for the bytes on each page and the access percentage associated with each page leads to the following:

$$120\,500 * .40 + 80\,500 * .30 + 60\,500 * .30 = 90\,500 \text{ bytes}$$

Thus, each hit on the organization's Web server will result in a requirement to transmit 90 500 bytes of data from the server to the Internet via the WAN connection to the Internet access provider.

Hit estimation

Perhaps the most difficult estimate that you must make when determining an appropriate WAN operating rate is the number of hits that you expect to occur during the busiest hour of the day. The reason that this estimate can have a high degree of variability is due to the fact that access to your organization's Web server can be based on a large number of variables, with many of those variables beyond the control of your organization. For example, although you can control advertising of your Web's URL in trade publications, it may be difficult, if not impossible, to inhibit robot search engines from visiting your site, retrieving each page available for public access on your server, and indexing the contents of your server's Web pages. Once this index is placed into the database of a search engine, access to your Web server can result from persons initiating a Web search using Yahoo, Lycos, Alta Vista or a similar search engine. Unfortunately, due to limitations associated with many search engines, forward references to your organization's Web server may not be relevant and can alter the distribution of page hits, as many persons on viewing your home page may click on the back button to return to the list of search matches provided by a search engine query and select a different match. For example, if your organization were a tire distributor named Roosevelt Tires, many Web search engines would return your home page URL in response to a search for Roosevelt, even though the person was searching for references to one of the presidents and not for an automobile tire distributor.

Although estimating the number of hits that will occur during the busy hour can be a difficult task, many Internet access providers can furnish statistics that may be applicable to use by your organization. A major exception to using such average statistics is if your organization is placing controversial or highly desirable information on the Web server, such as the results of Olympic Events as they occur, a swimsuit calendar, or some mixture between a Playboy bunny and a Coca Cola Polar Bear. Otherwise, the information concerning busy hour hits that your Internet access provider supplies can be considered to represent a reasonable level of activity that will materialize.

Returning to our estimation process, let us assume that your organization can expect 660 hits during the busy hour. Note that although this hit activity may appear to be low in comparison to the tens or hundreds of thousands of hits reported by well known URLs representing popular Web server sites, during a 24-hour period you are configuring the operating rate of the WAN connection to support 24×660 or $15\,840$ hits based on a busy hour hit rate of 660. According to statistics published by several Internet access providers, during 1996 the average number of hits per Web site when the top 100 sites are excluded is under 5000 per day. Thus, if your organization is the typical business, college or government agency, you may be able to use a lower wide area network operating rate than will be determined by this example.

Computing the WAN operating rate

Once you have determined the number of hits that you will support during the busy hour and the average number of bytes that will be transmitted in response to a hit, you can compute an initial WAN operating rate. The reason the computation will produce an initial WAN operating rate is the fact that it does not consider the effect of traffic on the LAN that the Web server is connected to. Thus, when we consider LAN traffic we may be required to adjust the WAN operating rate we will compute.

Based on the composition of our Web server, we previously computed that each hit will result in the server transmitting 90 500 bytes. As we expect 660 hits during the busy hour, we must initially size the WAN connection to support 660 hits per hour, with each hit resulting in the server transmitting an average of 90 500 bytes to the Internet. Thus, the initial WAN operating rate is computed as follows:

$$\text{WAN operating rate} = \frac{\text{average bytes/page} \times \text{hits/busy hour}}{60\,\text{min/hour} \times 60\,\text{s/min}}$$

Substituting, we obtain

$$\frac{660\,\text{hits/hour} \times 90\,500\,\text{bytes/hit} \times 8\,\text{bits/byte}}{60\,\text{min/hour} \times 60\,\text{s/min}} = 132\,733\,\text{bps}$$

Considering LAN bandwidth

Based on the preceding computations, you would be tempted to order a 192 Kbps fractional T1 as the WAN connection to the Internet access provider. You would be tempted to select the 192 Kbps fractional T1 operating rate because the next lower fraction of service, 128 Kbps, would not provide a sufficient operating rate to accommodate the computed busy hour transmission requirement for 132 733 bps. However, before ordering the fractional T1 line, you should consider the average bandwidth the Web server can obtain on the LAN that it is connected to. If the average bandwidth exceeds the computed WAN operating rate, the LAN will not represent a bottleneck that should be modified. Otherwise, if the average LAN bandwidth obtainable by the Web server is less than the computed WAN operating rate, the local area network will on occasion function as a bottleneck, impeding access via the WAN to the Web server. This means that regardless of any increase in the operating rate of the wide area network connection, the ability of users to access your organization's Web server will at times be restricted by local traffic on your LAN. If this situation should occur, you must then consider segmenting the LAN, creating a separate LAN for the Web server, migrating to a higher speed technology, or performing a network adjustment to remove the effect of a portion of local LAN traffic functioning as a bottleneck to the Web server.

To illustrate the computations involved in analyzing the effect of local traffic, let us assume that the LAN illustrated in Figure 7.2 is a 10 Mbps 10BASE-T network that supports 23 workstations and one file server in addition to the Web server, resulting in a total of 25 stations on the network. This means that on average each network device will obtain access to 1/25th of the bandwidth of the LAN or 400 000 bps (10 Mbps/25). However, the bandwidth of the LAN does not represent the actual data transfer that a network station can obtain. This is because the access protocol of the network will limit the achievable bandwidth to a percentage of the statistical average. For example, on an Ethernet LAN which uses the Carrier Sense Multiple Access Collision Detection (CSMA/CD), access protocol collisions will occur when two stations listen to the network and, noting an absence of transmission, attempt to transmit a frame at or near the same time. When a collision occurs, a jam signal is transmitted by the first station that detects the high voltage resulting from the collision, causing each station with data to transmit to invoke a random exponential backoff algorithm. This algorithm generates a period of time during which the network station delays attempting a retransmission; however, the frequencies of collisions, jams and the invocation of backoff algorithms increase as network utilization increases. For an Ethernet network, network utilization beyond a 60% level can result in significant degradation of performance which can serve as a cap on achievable transmission throughput. Thus, the average bandwidth of 400 000 bps previously computed should be multiplied by 60% to obtain a more realistic level of average available bandwidth obtainable by each station on the LAN, including the Web server. Thus, in this example, the Web server will obtain on average $400\,000 \times .6$ or 240 000 bps of LAN bandwidth. As the average LAN bandwidth obtainable by the Web server exceeds the computed WAN operating rate, no adjustment is required to the LAN. If the Web server were connected to a Token-Ring LAN, the average bandwidth of 400 000 bps should be multiplied by 75% as a Token-Ring LAN does not have its performance serious degraded until network utilization exceeds 75%.

Web page adjustments

As indicated in this section, you must consider the LAN bandwidth obtained by the Web server as well as the WAN operating rate to effectively select a wide area network connection method to an Internet access provider. When computing the WAN operating rate, it is important to note that that rate depends on the number of hits you expect to occur during the busy hour, the storage in bytes required to hold each page which represents data that has to be transmitted in response to a page hit, and the distribution of hits on the pages that are placed on your Web server. Although the first and third factors are obtained by an estimation process, you have a high degree of control over the composition of your server's Web pages that can be used as an effective tool in adjusting the ability of a WAN connection to support the estimated number of hits that you expect to occur during the busy hour. As initial access to your Web server is via its home page, that page will have the highest distribution of hits on your server. Thus, if your estimate of busy hour traffic is low, you can increase the ability of a selected wide area network operating rate to support additional hits by reducing the transmission associated with each home page hit. To do so you can consider replacing GIF images by their equivalent JPEG images that require less storage, cropping images to reduce their data storage requirements, or eliminating all or some images on the home page.

When you examine the composition of your Web pages it is important to remember the characteristics of HTTP. That is, each graphic retrieved by a client–server session requires a separate session. Thus, the initial handshake via the Internet, including synchronization between a browser and Web server software is repeated for each graphic on a Web page. Before you decide to place ten small graphic buttons on a page, use a dozen icons to spice up your page, or consider a collage of photographs, think twice about the remote user accessing your page. Will they be using a relatively low speed modem connection and have to wait a relatively long period of time that may result in their impatience breaking the connection to your servers? Even if they are connected to the Internet by a relatively high speed connection, are the results of the multiple sessions worth a short wait? In effect, become a devil's advocate when you view the composition of your organization's Web server and eliminate or combine unnecessary or only partially necessary images.

7.2 INTRANET CONNECTIVITY

The application of Internet technology to private networks is considered to represent an intranet. Although this term is relatively new, it simply refers to the potential use of Web servers and browsers as well as Gopher, Archie, Veronica, Finger, Whois, and other TCP/IP applications. In this section we will focus our attention on Web browser to Web server communications on a corporate intranet, because the primary purpose of this chapter is to assist readers in selecting an appropriate Web server connection rate.

Overview

As an intranet represents the application of TCP/IP technology to an organization's internal private network, the selection of an appropriate Web server connection rate can be very different from that previously discussed for an Internet. This is because an internal network may not require a connection to the Internet and may be structured as a grouping of local area networks connected in some manner to form a private network. Thus, any discussion of Web server connection rates with respect to an intranet requires a discussion of various intranet network configurations.

Local LAN connection

When a Web server is connected to a local area network that has no other connections, the server's connection is normally at the LAN's operating rate. Although you may be tempted to ignore the effect of the composition of Web pages when servers are connected to a LAN and other network users accessing servers operate at LAN operating rates, to do so can result in network utilization problems. To illustrate this, assume that an existing network has a 55–60% level of utilization. Even if the use of a Web server as a help desk adds only 3–5% utilization to the network, it would then saturate the LAN. This is because Ethernet performance begins to seriously degrade at utilization levels beyond 50–55% utilization.

Through the use of an intelligent switching hub, you can consider the use of a higher speed network connection or the use of what is referred to as a 'fat pipe' to facilitate browser to Web

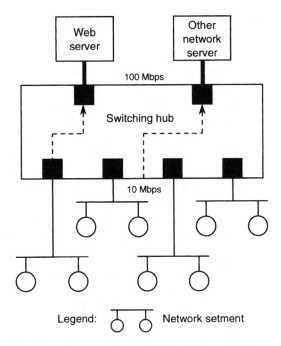

Figure 7.3   Through the use of a switching hub multiple connections can be supported

server connectivity while minimizing the effect of that connectivity on other network activity.

Figure 7.3 illustrates the use of an intelligent LAN switch high-speed port to connect a Web server to a corporate intranet formed through the connection of a number of network segments to the switch and the use of TCP/IP applications. In this example a high-speed 100 Mbps port is used to connect the Web server to the switch. Although the LAN segments may only operate at 10 Mbps, the use of a 100 Mbps server connection enables the server to fulfill requests faster, allowing the server to service a second request when it would otherwise be transmitting the response to a previous request. If the switch supports a 'fat pipe' instead of a high-speed port, this means that several connections to the server would be treated as an entity to provide a similar effect to the use of one high-speed connection. In addition to providing a high-speed connection, the use of a switching hub enables multiple simultaneous communications to occur through the switch. This is indicated in Figure 7.3, which shows one connection from one network segment to the Web server, whereas a second connection is occurring between a station located on a different segment and a different network server.

Internet access

If your intranet is connected to the Internet you will use some type of wide area network transmission facility to connect the LAN on which the Web server resides to the Internet. In this situation you can consider the previously described method in section 7.1 to determine an appropriate WAN operating rate.

The selection of an appropriate wide area network operating rate to connect your organization's Web server to the Internet or a corporate intranet depends on three key factors, of which two (the expected number of hits during the busy hour and the distribution of hits per server page) can only be estimated. Although this means that the ability of the WAN operating rate to service expected traffic will only be as good as two traffic-related estimates, by planning ahead to adjust the third factor, the data storage on your server's home page, you obtain the flexibility to alter the ability of a selected WAN operating rate to support additional hits during the busy hour. Thus, the methodology presented in this chapter provides you with the tools to select an appropriate WAN operating rate as well as a methodology to obtain flexibility in adjusting the ability of the selected operating rate to support additional hits. By following the methodology presented in this chapter, you will be able to remove a large degree of the guesswork associated with connecting a Web server to the Internet, as well as the ability to rapidly adjust the capacity of a wide area network connection to support additional Web server hits if such an adjustment should become necessary.

IMAGE TRANSMISSION TECHNIQUES

The significant increase in the use of electronic mail to transport text, digitized voice in the form of audio clips, images and digitized video clips can result in several interesting problems when the originator and recipient of a message are located on different electronic mail networks. Those problems include the inability of many gateways to support the transfer of binary files, whereas other gateways restrict the size of files transferred between e-mail systems. To overcome these problems, several software developers introduced software solutions that incorporate techniques that can be used to overcome this apparent interoperability. In this chapter we will examine the rationale for using these techniques as well as to look at the actual use of several techniques to transfer binary files between different electronic mail networks. As interoperability problems primarily result from the operation of e-mail gateways, a good place to commence our examination of image transmission techniques is by first discussing e-mail gateway operations.

8.1 E-MAIL GATEWAY OPERATIONS

An electronic mail gateway can be considered to represent a conversion device which transfers electronic mail messages between networks connected by those gateways. Some gateways convert a message transmitted as a file under the format established by one e-mail program into a format required by a second e-mail program. Other gateways simply transfer files from one network to another. For both situations, gateways operate according to a programmed set of rules that may restrict the type and size of files transmitted from one network to another.

Interoperability problem evolution

When the Internet was originally constructed, e-mail gateways were developed to provide electronic mail interoperability between different networks on this network of networks. At that time there was no digitized voice, the use of images was limited, video clips were not being transmitted via e-mail, and the vast majority of electronic mail messages consisted of plain ASCII text. Although this restriction did not adversely impact the transmission of electronic mail messages during the early 1970s, by the late 1970s and early 1980s it was well recognized that a mechanism was required to transmit binary files as attachments to electronic mail messages. The reason for this was the significant increase in the use of wordprocessing programs that embedded 8-bit codes in wordprocessing documents to indicate line positioning, font size and other text attributes. Another reason for the driving need to transfer binary files between e-mail systems was a significant increase in the use of a variety of images to include clip art in word-processing documents. Regardless of whether a wordprocessing document contains clip art, images, fonts or other text attributes, the result is a binary file that cannot be transferred between many electronic mail systems.

Examining e-mail interoperability

To illustrate an e-mail interoperability problem, consider Figure 8.1 which illustrates a Word Perfect document containing an image appropriately placed on the page. If you save the Word Perfect document shown in Figure 8.1, the program will prompt you to determine whether or not the image should be included in the saved document. Unless you simply save the document as an ASCII file and lose both the image and the large bold font that refers to the name of this author's pet, from the perspective of transmitting the document via e-mail it does not matter whether or not you include the image. This is because the document, with or without the image, will be represented as a binary file which many gateways cannot support. Of course, if you save the file without the image, the resulting size of the file and its transmission time will be significantly less than if the file with the image was saved. In addition, due to the size restrictions associated with many electronic mail systems, you may be able to transmit a small binary file, whereas the e-mail system may refuse to accept or transfer a larger file containing one or more

Figure 8.1 A Word Perfect document that must be saved and transmitted as a binary file due to its use of text attributes and inclusion of an image

images. As we now have a file to transmit, let us attempt to do so, using CompuServe in an attempt to transfer the file to a subscriber of MCI Mail.

Using CompuServe

To illustrate a typical problem that you can expect to encounter when transmitting large binary files between different electronic mail networks, this author first attempted to transmit an electronic message with the document illustrated in Figure 8.1 used as an attachment to the message. As the Word Perfect file is slightly over 1.1 Mbytes in size due to the inclusion of an image, this file can be considered to represent a relatively large e-mail message, especially when surveys indicate that the average e-mail is under 2500 characters, including its headers.

Figure 8.2 illustrates the use of the CompuServe Create Mail dialog box to create a short message, after which the Attach

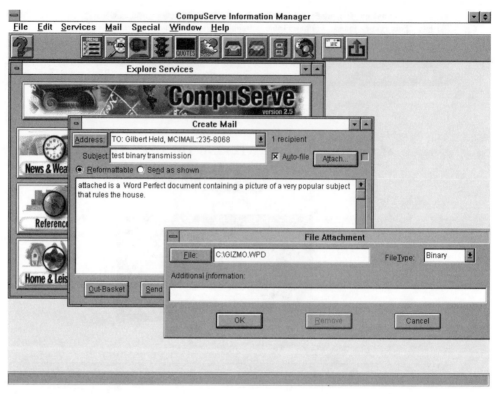

Figure 8.2 Using CompuServe in an attempt to transmit a large binary file to a subscriber on MCI Mail

button was selected. That operation resulted in the display of a dialog box labeled File Attachment which is shown superimposed on the dialog box labeled Create Mail. Note that the file GIZMO.WPD was selected, which represents the name of the file used to store the document shown in Figure 8.1, and the File Type in the File Attachment dialog box was set to Binary.

Figure 8.3 illustrates the initial attempt to transfer the previously created short message that includes the attached binary file. Unfortunately, CompuServe apparently considers the size of the file to be excessive and quickly terminates the session, with only the message and not the attached binary file being transmitted to the message recipient. Thus, it appears that a file size of 1.14 Mbytes exceeds the capacity of CompuServe.

To illustrate that it is possible to transfer certain binary files between different e-mail systems, let us return to the use of Word Perfect and go to the opposite extreme and create a rather small binary file. Figure 8.4 illustrates the use of Word Perfect to create the message 'This is a test' in a large bold font, using a font size

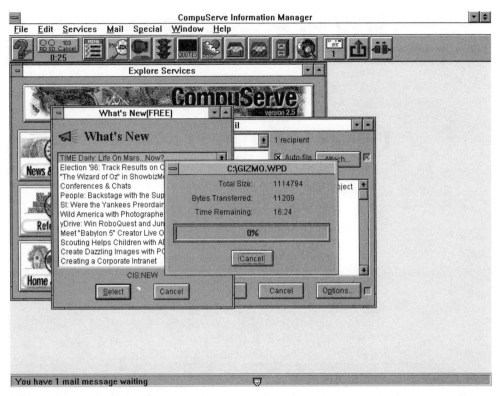

Figure 8.3 An attempt to transfer a very large binary file will result in CompuServe eventually terminating the session

of 36. Saving this most creative message using the file name TEST results in Word Perfect adding the file extension .WPD. The total size of this file is 1229 bytes, although the message consists of 14 characters including spaces between the four words in the message. The additional bytes include Word Perfect generated binary codes that define the text attributes including their font size as well as their positioning on the document. In addition, Word Perfect documents include information about the default printer the user used, which adds to the resulting file size. As we want to determine if we can use CompuServe to transmit a smaller size binary file to a subscriber of MCI Mail, let us attempt to do so.

Figure 8.5 illustrates the use of the CompuServe dialog box labeled Create Mail to create the message 'attached is a binary file' which will serve as a prompt to notify the recipient that a binary file is attached to the message. The dialog box labeled File Attachment, which is partially superimposed on the previously mentioned dialog box, illustrates the selection of the file

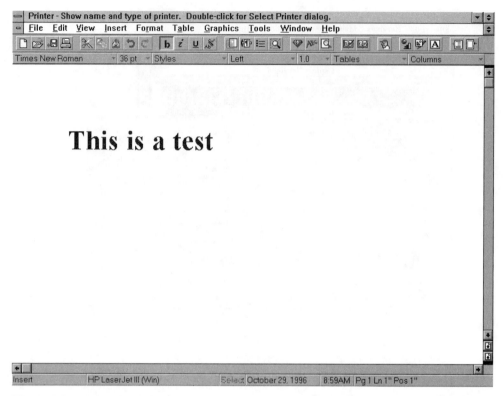

Figure 8.4   Using Word Perfect to create a small binary file as a mechanism to test the transfer of binary files between different electronic mail systems

TEST.WPD from a disk in drive A and the selection of binary from the file type selection button. Once we have clicked on the OK button on the dialog box labeled File Attachment and the Send Now button on the dialog box labeled Create Mail, CompuServe will accept this smaller sized file. Within a short period of time this author can log onto his MCI Mail account and scan his inbox. The result of that scan is shown in Figure 8.6. In examining the contents of Figure 8.6, note that two messages were received. The first message contains the text 'attached is a binary file,' whereas the second message actually contains the binary file that was attached to the CompuServe electronic mail message illustrated in Figure 8.5.

In examining Figure 8.6 you will note that the size of message number 18 is shown as 29 bytes. That message represents the original CompuServe created e-mail message without the binary attachment. However, when you start to download the file, you will observe MCI Mail displaying the Message 'Filesize 499 characters,' as we will shortly note. Thus, a logical question is where

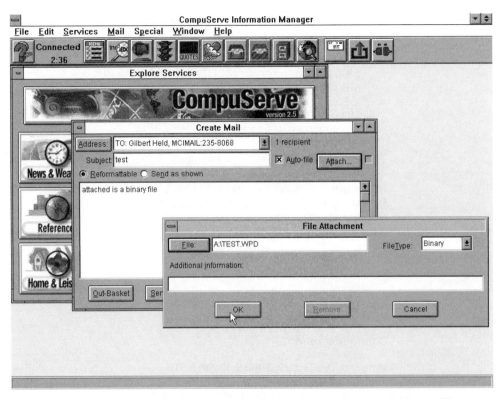

Figure 8.5 Using CompuServe a second time to transmit a small binary file as an attachment to an e-mail message destined to a subscriber on MCI Mail

do the extra characters come from? In actuality, those characters represent an x.400 e-mail header added to the message. Similarly, the binary portion of the message that represents the attached file to the original message will also have a header. This is illustrated in Figure 8.7 which shows the results obtained from reading the two messages received by MCI Mail. Note that

```
Command: scan 18–19

    2 messages in the most recent scan

No. Posted       From            Subject    Size
18  Oct 29 09: 12  X.400 Originator   test     29
19  Oct 29 09: 12  X.400 Originator   test    1229

Command: →
```

Figure 8.6 Scanning the inbox on MCI Mail indicates that the Compu-Serve e-mail and file attachment are received as two separate files

read 18

Date   Tue Oct 29, 1996 9:12 am EST
From: X.400 Originator
        EMS: CompuServe / MCI ID: 592-7515
        MBX: S=
        MBX: C=US
        MBX: P=CSMAIL
        MBX: DDA=ID=73233.1335

TO: * Gilbert Held / MCI ID: 235-8068

Application message id: 961029140658 73233.1335 FHO48-1
Posted date: TUE OCT 29, 1996 2:06 pm GMT
Importance: Normal
Grade of Delivery: Normal

Subject: test

attached is a binary file

Command: read 19

Date: Tue Oct 29, 1996 9:12 am EST
From: X.400 Originator
        EMS: CompuServe / MCI ID: 592-7515
        MBX: S=
        MBX: C=US
        MBX: P=CSMAIL
        MBX: DDA=ID=73233.1335

TO: * Gilbert Held / MCD ID: 235-8068

Application message id: 961029140703 73233.1335 FHO48-2
Posted date: TUE OCT 29, 1966 2:07 pm GMT
Importance : Normal
Grade of Delivery: Normal

Subject: test

A binary file entitled: GRAPHICS. is included with this message and
can be downloaded via your communications software. Please type HELP
DOWNLOAD MESSAGE at the Command: prompt for instructions

Command: →

Figure 8.7 The headers for the two messages received by MCI Mail.
Each header conatins x.400 information and adds to the size of the
received file

the header for the binary file does not actually include the file attached to the original message for listing purposes. Instead, it informs the recipient that a binary file is included with the message and can be downloaded via communications software. Perhaps a better explanation would be for MCI Mail to prompt the recipient to use a binary file transfer protocol to download the file.

Figure 8.8 illustrates the use of ProcommPlus to download the binary attached file that was originally attached to the Compu-Serve message. In this example, the ZMODEM protocol is shown being used. Although the file size in the ZMODEM file transfer protocol status box shown in Figure 8.8 is indicated to be 499 bytes, that actually represents the first of two files associated with the CompuServe binary file attachment that will be transmitted by MCI Mail. As previously noted, the first file contains an x.400 electronic mail header whereas serves as a prefix to the message, whereas a second file will contain the actual binary file that was transmitted as the attachment to the CompuServe electronic mail message. Thus, by carefully examining the background portion of the screen display shown in Figure 8.8, you will note that the MCI

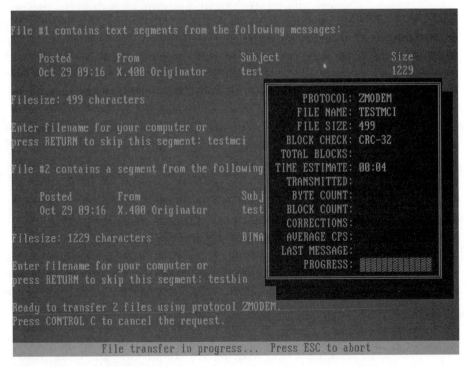

Figure 8.8 Using ZMODEM to download the header associated with the binary files followed by that file

Mail system prompts the user to enter two file names for the transfer, with the foreground ZMODEM file transfer protocol status box illustrating the beginning of the first file transfer that contains x.400 header information associated with the original message.

Although we were able to transfer the contents of a relatively small binary file, as previously noted we could not transmit a relatively long binary file. This indicates a common problem associated with many electronic mail systems: the inability to support files larger than a certain length. As the original file containing a picture of the author's dog was approximately 1.14 Mbytes, it is obvious that CompuServe does not support e-mail transfers of that magnitude. Thus, a key question that you will require an answer to when attempting to transmit most types of image files via electronic mail is the maximum length of an electronic mail message supported by the originating and receiving e-mail systems. Then you must structure your e-mail message to obtain a maximum file size that is less than or equal to the lowest file size supported by the originating or receiving network.

Today most electronic mail systems commonly support a maximum e-mail file length of 65 536 bytes. In certain situations where disparate e-mail systems are directly connected to one another, much larger file transfers may be possible. In fact, several additional transfers from CompuServe to MCI Mail resulted in files as large as 280 Kbytes being e-mailed between the two systems. When your file size exceeds the maximum length supported by one or more e-mail systems, or when you cannot transfer binary files between different e-mail systems, you must then consider the use of one or more software-based techniques to obtain interoperability. As these techniques are based on the encoding of 8-bit binary characters into 7-bit ASCII, we will examine them as an entity in the next section in this chapter.

8.2 ENCODING/DECODING BINARY FILES

The development of a variety of electronic mail systems during the late 1960s and early 1970s was accompanied by the recognition of the fact that a method was required to permit binary files to flow through gateways that were restricted to supporting 7-bit ASCII characters. One of the earliest techniques developed was UUencoding, which has its roots in Unix and whose UU represents the term Unix-to-Unix. Other coding techniques that have gained a moderate to high level of acceptance include XXencoding and Base 64 encoding. As many software techniques designed to split large binary files into smaller entities for transmission between different e-mail systems are based on the use of one or more of these encoding methods, we will first focus our attention on their operation in this section. Once this has been accomplished we will then examine how these techniques have been implemented into e-mail communications systems as well as stand-alone software products to provide e-mail users with the ability to transfer both small and large binary files between different electronic mail systems.

UUencoding

UUencoding represents a technique in which eight-bit characters are converted into printable characters for transmission through and between communications devices that cannot support eight-bit character transfers. To accomplish this, UUencoding first groups characters in the original message together to form tri-byte groups of 24-bits. Each tri-byte group is then converted into four six-bit characters, with a space character with a decimal value of 32 added to each six-bit character. The result of this character addition process is to ensure that every encoded 6-bit character has a value above decimal 32. This precludes the mapping process converting encoded characters to values below 32 which are normally reserved for communications functions, and whose interpretation could result in unintended actions from gateways that interpret the values of characters transmitted between e-mail systems. As a result of the previously described encoding process, binary characters are mapped into the following character set:

'!"#$%&'()*+,−ˆ/0123456789:;<=>@ABC...XYZ[\]ˆ_

In examining this encoding technique, you will note the absence of spaces in the encoded character set. In actuality, spaces are mapped into back-quote (') characters whose value is decimal 96. The reason for this mapping is the fact that many gateways remove trailing spaces on lines, an event that could play havoc with the decoding process at the recipient's end if the encoding process resulted in spaces at the end of one or more encoded lines. Concerning decoding, the second part of the pair of programs is commonly referred to as UUdecode. This program decodes or restores a UUencoded file to its original form.

XXencoding

A more recently developed but less popular method employed to convert 8-bit binary information into printable ASCII characters suitable for transmission through gateways and other code-sensitive communications devices is referred to as XXencoding.

XXencoding is similar to UUencoding in that it converts groups of tri-bytes into four six-bit characters. The key difference between the two encoding methods is the character set used for the translation process. XXencoding uses the following character set:

$$+-01 \dots 89ABC \dots XYZabc \dots xyz$$

In comparing mapped character sets used by UUencoding and XXencoding, you will note that the latter encoding method is restricted to generating the digits, upper and lower case letters of the alphabet, and the plus (+) and minus (−) signs. The elimination of many special characters from the mapped character set can alleviate translation problems associated with the transfer of messages between e-mail systems that are transported in different character codes, such as ASCII, extended ASCII and EBCDIC.

Base 64

A third commonly used encoding scheme is also based on the regrouping of three characters into four six-bit characters to generate a new 64 character set. Referred to as Base 64 encoding, this technique is incorporated into the Multipurpose Internet Mail Extensions (MIME) standard. Under Base 64 encoding, each group of 24 bits from three input bytes is first converted into four six-bit characters. Next, each six-bit character is prefixed with the binary value 01, resulting in each character being in the decimal range 64 to 127. Although Base 64 encoding is gaining in popularity due to the fact that it is the only standardized encoding technique, its use is far from universal. Although Base 64 encoded files can be transmitted between most e-mail systems, if the recipient's system does not include a Base 64 decoder that automatically restores the file to its original composition, that person must either acquire a stand-alone decoder or request the originator to resend the file using an encoding mechanism the recipient's decoding mechanism supports. As UUencoding remains as one of the most commonly used methods to transmit binary files between different electronic mail systems, we will examine the operation of two programs that can be used to transmit images as attachments to electronic mail. In doing so, we will also note how the use of an appropriate UUencoding program will subdivide a large file into a series of smaller files that alleviate the file size restrictions associated with many electronic mail systems.

Using UUENCODE/DECODE

UUENCODE and UUDECODE represent a program pair of encoding and decoding programs developed by Richard Marks of Bryn Mawr, PA. Dr. Marks originally developed this program pair in 1987 and has significantly updated his original programs, adding support for other encoding methods, the automatic separation of files over 64 Kbytes into multiple encoded files and their recombination, as well as other features. The basic program pair is included on the CD-ROM accompanying this book and is described in additional detail in Chapter 9.

Optional parameter switches

Figure 8.9 illustrates the UUENCODE help screen which is displayed when the program is invoked without any parameters in its command line. As indicated in Figure 8.9, the command line

```
C:\UU>uuencode

UU-ENCODE 95 (v40) FOR PC. by Richard Marks

Usage: uuencode [-clshouxt] <file-to-encode> [<output-file>]
       -c : do not create any checksums.
       -l : put checksum on every line.
       -s : do not split output file into sections.
       -s nnn : section contains nnn lines (default=950 lines).
       -h nnn : leave room in first file for nnn header text lines.
       -o : write to standard output.
       -u : create unix format file (default is MS-DOS)
       -x : encode using XXDECODE characters.
          upper case X, also default to .XXE extension.
       -6 : encode using Mime compatible base 64 form
       -t : put character mapping table into output

-or- see UUSER.TXT for more info

C:\UU>
```

Figure 8.9 The UUencode help screen

entry includes optional parameter switch options followed by the name of the file to encode and the name of the output file. The output file name is optional, and if not included the program will use the input file name as the output, replacing any filename extension with the extension .UUE.

In examining the program switch options listed in Figure 8.9, note that the −s option disables the automatic splitting of a file that exceeds 64 Kbytes. Thus, if the source and destination e-mail systems and the gateway linking the two support the transmission of relatively large files, you can use the −s option or the −s nnn option, with the latter used to split the encoded file into nnn line sections, with each line containing 45 or fewer encoded characters.

Operation

To illustrate the use of UUENCODE, let us again use the small Word Perfect document that we previously created, this time UUencoding the file prior to transmitting it as a binary attachment. Although we will be encoding a binary file that represents a wordprocessing text document, the use of UUENCODE is applicable to any type of binary file, including documents that contain images or images stored separately in a particular file format. In using UUENCODE we would enter the following command to convert the previously created Word Perfect binary file into a UUencoded file.

UUENCODE TEST.WPD TEST.UUE

Figure 8.10 illustrates the contents of the resulting UUencoded file. Note that the top line in the file indicates that the conversion required one file display, the name of the original file, the program name and version number, and the initials of the program author. That line is followed by a line with the keyword 'begin.' This denotes that all lines following the line above the keyword 'end' contain the actual encoded data. The number 644 in the line commencing with the word 'begin' is the file mode and it represents a carryover from the Unix *chmod* command. In the UUENCODE program developed by Dr. Marks, the file mode value is fixed at 644. Between the lines labeled 'begin' and 'end,' you will note that most lines begin with the letter M. Most lines in a UUencoded file have 45 encoded characters. The character M is decimal 77. As each encoded character has 32 added to it, subtracting the previous addition from 77 results in the line character count of 45. The last line in an encoded file is a line with zero encoded

section 1 of 1 of file test.wpd <uuencode 95 (v40) by R.E.M.>

```
begin 644 test.wpd
M_U=00RH$$$'"'!"@(!'""'@4"#'##-!"'"'('&!/_0:H_@(CSO$Q&@&7$$DDJ>J
M1?I$$B$B<,,^=!2!D(@)ZH&J&/*$$)%!D$*1-OG,J(PA'E]O-6?E,Gj,J(>-8[NI
M()NKj&H.L$-=A1W$P$:.?D1'>CS[.JT\WVK,?XA$GH'/&E=/?[')+3A^U7">
M@@TA(8N[Cl2OA]<3CDX[.!O^H;6K<m]C0?247,GX9>8.Q:-GB[)QE'K=5PDH5
M_9(;Cl6X[O<(SB0Z_?6?"P'GK/'m9L?R0XL$0%_5!(UE]&X7\D/E5BK3^+#1
M!TV!<2CJ.$(E4('W0#UWDZ"G>OWKXm]D$CQ@("KL!"'$IC:FVD..Y/%H2KL">_
M-W(PL3*/MZ1?Y/&WSEYVDV'O:;_JSKg.V"8;LI>0**UC8k;8;H=ZA*$$Mk'
MP@H?F5@]V&'3VXN38;_30'F?1X:C,<['->(JE480$)R3!1_)O$'/E%W\S"!
M(((($1?1?)'\H,4KZ-(7W;3T#Q@._+X/)?7MS\,%L)-3,+3LZGk\1\MQ#>O@B6Jk'Z8M
MBN0M62D['6@m0'AK<)$)T[2SQ;WDD==;0XGI?Q4,N:N1L:ANX&<<$$F20LSB.G
MZT?K:U:@9]D\6V-%EE''P5#J0CN#lSJSkZ9;[Q@44IMF:T:[TEH22SU/@>5H'
MH+'OF5LAAJ5&T3Xj[Z85K'<"''@''''''''''''@C'0'''L!''!P''@''
M'%4#'''3@''L#'''))0$''''&''''R0,''LP''@''''@''#/'P''''<!
M''''$''''/<#'''(-'$'''4'''!P0''@''O''\''';!'''')A('%''
M('!,'&$$'(2@!E'0('''!]'20''E'0'('''$D'20''''''''''''''''''''
M''''''''''''''''''''''$$$$04$$$-,,-4U3'''''L'2P!+''L
M'2P!+''$$L'2P!,''''''''''''''''''''''''''''
M'''''''''''''''''''''''''''''''''
M'''''''''''''''''''''''''''''''''
M'''''''''''''''''''''''''!>',<X*'#6'L,,.0@'
M'!$)''''6@'+0''+%#&'5'II'&T'90!S'''3@!E''<''('!2'&\''';!'A'&X'
M('!2'&4''9P!U'&P'80!R'''0'''%''@''0'''0'*''''''''
M''''''''''''!$$@(')''A''H''H@''''!''(@''''''','P!''9VX@
M''''''''''''''''''S?!IX'$$'''''''''J4*$$$'#'00'P'''$0
M''-W++''PL'P''?Y,,0$$'(''=''=!8'''P#4\0'#'#-!''''.'#''#''
M'''((P#4\PSS'0'''/$
'''''''''''''''''''''
```

end
sum-r/size 55767/1720 section (from "begin" to "end")
sum-r/size 19579/1229 entire input file
→

Figure 8.10   Results obtained from the UUencoding of the file test.wpd

characters. Thus, you will note the back-quote (') character above the line labeled 'end.' At the end of each section, the program (unless a suppression option is used), inserts lines containing checksum information for the section between 'begin' and 'end' and the entire input file, providing a mechanism for UUDECODE to perform an integrity check on the encoded file as it performs its decoding operation to reconstruct the original file. This checksum capability is not included in many similar implementations of UUencoding and decoding programs, and it can be a most valuable feature that warrants the use of this stand-alone pair of programs instead of similar programs either automatically invoked by some e-mail systems when binary files are transferred, or provided as an additional feature incorporated into other types of programs.

To illustrate the transmission of the previously UUencoded file, that file will be used as an attachment to a Word Perfect Office e-mail message. Figure 8.11 illustrates the creation of the short message 'uuencoded file attached,' with the file test.uue attached to the message. As indicated in the Word Perfect Mail window, the message will be transmitted to this author's MCI Mail account via an Internet connection.

Transmitting the message from an organization's Word Perfect Office e-mail system to an MCI Mail subscriber results in the message and its attached file flowing through at least one e-mail gateway. Thus, this e-mail transmission will test the ability of a UUencoded file to flow between Word Perfect Office and MCI Mail via the Internet. Although Word Perfect Office has an automatic encoding capability that operates on the contents of binary files, the encoding method supported differs between versions of Word Perfect Office. In addition, the recipient may not have a decoding program that is compatible with the originating encoder. Thus, often it may be preferable to use a stand-alone program to encode binary attachments prior to their transmission between different

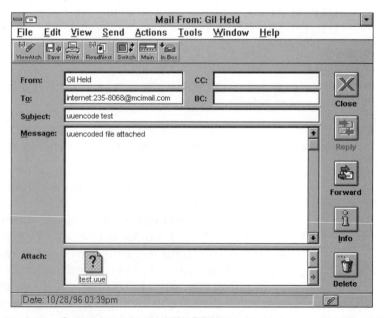

Figure 8.11 Creating a short UUENCODE test message using Word Perfect Office, with the UUencoded file attached to the message

e-mail systems, even if the originating e-mail system supports one or more of the encoding methods previously discussed.

Unlike many binary files that cannot be read unless the application used to create the file or an application that supports multiple file formats is used, test.uue is an ASCII file that can be read by almost all wordprocessing programs and text editors. Figure 8.12 illustrates the partial contents of the file that was displayed when the author double-clicked on the test.uue file icon shown in Figure 8.11. If you compare the contents of Figure 8.12 to Figure 8.10, you will note that the files are identical and the file test.uue attached to the e-mail message indeed represents the UUencoded file.

As we are probably anxious to view the message received on MCI Mail, let us do so. Figure 8.13 illustrates the display of the previously created Word Perfect Office message and attached UUencoded file which resulted in the message and the attached file being received as one. Note that the received message includes the

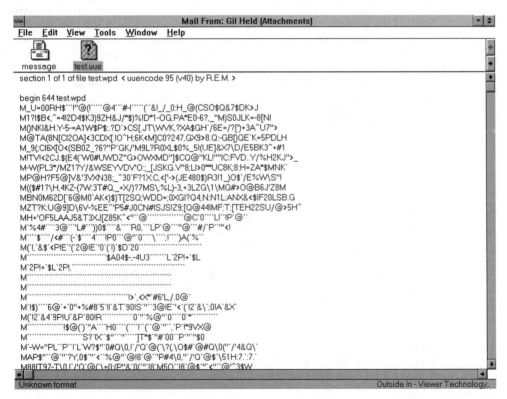

Figure 8.12 Viewing the partial contents of the UUencoded file test.uue attached to the previously created message. The encoded file was created using a stand-alone UUencoder

Date: Mon Oct 28, 1996 6:11 pm EST
From: Gil Held
     EMS: INTERNET / MCI ID: 376-5414
     MBX: GXEIELD@opm.gov

TO: * Gilbert Held / MCI ID 235-8068
Subject: uuencode test

uuencoded file attached

------------------TEST.UUE follows ---------------- -----
section I of I of file test.wpd < uuencode 95 (v40) by R.E.M. >

```
begin 644 test.wpd
M_U==00RH$"$&$"$G!"]@U[!.#.1#.$(5#.43/!=N#K/E/;:s)
M1?I$B<,,,^__^^s
M()NKI&H.Y-^..^^s
M@TA(8N[Cl20A]<3 CDX[.!O^H;6K<M]C0?247, GX9>8.Q :-GB[)QE`K=5PDLH
M_9(;Cl6X[O<(SB0Z_?6?"P`GK`M9L?R0RXL$0%^_5!(UE]&X7\D/E5BK3^+#l
M!TV!<2CJ$$(E4(`W0#UWDDZ"G>OWXM@.])$C@@K]L!V
M-WPL3*/Mz1?Y/&WSEYVDV`O:;/JSSKG.V"8;LI>O**UC8K8:H=ZA`$M.N 
M8H@(?F@@]V&'3 VXN3?;0`f>=(Jeq80$)r!>o0$/E%w\s"!
M(($$#1?\H,4KZ-(7W:3T#Q,)?7MS\;L&L)-3,-3 LZG\I\Mq#>O@b6J'Z8m
MBN0M62D['6@mOAK<)$]T[2SQWDD=;0XGi?Q4,N:N1L:ANX&<$IF20LSB.G
MZT?K:U@-9]D\6V-%EE^'P5#J0CN#iSJS!Z9;[q@44imf:T:[tEh22su/@>5h^
MHH+'OF5LAAT5&T3XJ[Z85K^<"`@`"@C`0'''L!'IP'@'
M'%4#'''3@''L#''))0$'''&''`R0,'`LP'@`''@''#/`P'''<!
M''$''''</#`'(-'$''4''!P0'''@'"O'''\'';;!''A);'%''
M('!,'&$$'<P!E'('2@!E'0'(!)'$'$'D'20''''''''''''''''''''''''''''
M'''''''''''''''''$$04#$,,,4U3''''''L'2P!)-'$L
M'2P!)-$$L'2P!,''''''''''''''''''''''''''''''''''''''''''''''
M''''''''''''''''''''''''''''''''''''''''''''''''''''''''''
M'''''''''''''''''''''''''''''!'>`,<X*(#l6'L,,0@`
M'!'$$')^@'+'0^+%$#85'5!L'''3@!E'''<'('l2'&\';;0!A'&X'
M('l2'&4'9P!U'&P'80!R'''''''''''0'''%'@'''0'''0'*'''''''''
M''''''''''!'$$'((@')'"A'''''A''H0'''''('('@''''''P'''*'9VX@
M''''''''''''''''S?!X''$'''''''')'*'$''#''''00''P''''!$$0
M'''-W=D"PL''''P!''''!'L'0'''0'.,!'/'''Q''@''''''/4&Q'
M'''A($'''''''?y,O0$''''''<''%'@''''l8!''P''P#%H:7':7'
M'M88!T97-T+4P'',!'/''''Q''''''''''''''''''''''''''''''''''''
M''''''''''''''''''''''''''''''''''''''''''''''''''''''''''''''''
.'''(P#4\PSS\0,'''/$
'
end
sum -r/size 55767/1720 section (from "begin" to "end")
sum -r/size 19579/1229 entire input file
→
```

Figure 8.13 The message and UUencoded file attachment appear as one file when retrieved from MCI Mail

checksum values which will allow the UUdecode program to verify the integrity of the received message. By simply downloading the file and using UUDECODE you can restore the binary file to its original format. For example, assume that you downloaded the full message shown in Figure 8.13 to the file TEST.UUE. If you then enter the command UUDECODE TEST.UUE the UUDECODE program would ignore the x.400 header and use the file name TEST.WPD in the line beginning with the keyword 'begin' to recreate the original binary Word Perfect document.

To illustrate potential compatibility problems that can occur when built-in encoders to e-mail programs are used, let us transfer the test.wpd document as a binary file using Word Perfect Office. Figure 8.14 illustrates the Word Perfect Office Mail window, indicating that the file created using Word Perfect will be sent as a binary attachment to an e-mail message to this author's MCI Mail account.

Figure 8.15 illustrates the resulting e-mail received on this author's MCI Mail account. Although the encoded portion of the file automatically produced by Word Perfect Office is very similar in appearance to the file created through the use of the UUENCODE program from Richard Marks, there are some differences between the two. Those differences are in the use of checksums and a header separator, both of which are generated by the UUENCODE program. Although UUDECODE can successfully

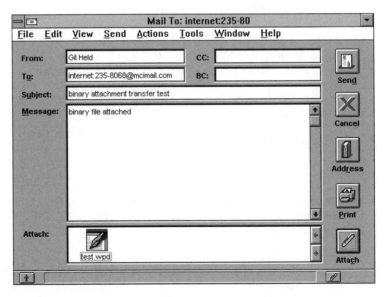

Figure 8.14 Using Word Perfect Office to transmit an attached binary file in the form of an attached Word Perfect document to an MCI Mail user

Date: Fri Nov 01, 1996 10:57 am EST
From: Gil Held
     EMS: INTERNET / MCI ID: 376-5414
     MBX: GIHELD@opm.gov

TO: * Gilbert Held / MCI ID: 235-8068
Subject: binary attachment transfer test

binary file attached

```
begin 644 TEST.WPD
M_U=00RH$"""!"@("!'""""'4'''#-!'""""('"&!./_0:H-_@_(CSO$Q&7$DK>J
M1?I$B"<,,^"=4!2D4$$K3)9ZH$&J/*$)%(D*1-OG,P%*E$*-6?",?,_M)S0JLK=-8[NI
M()NKI&H+ Y-5-=-=AAQ$K$$$P$$.?D$>CES[.JT\WVK,?XA$&(/6E=/?[`]+3A^U7">
M@_TH(8N[CI2OH]<3C+X[.!O^H^;;;<$]]$C0?247,GX9>8.Q.:-GB[])QE`K=5PDLH
M_9(;/;CI6x[o<(SBoz_?6?"P`gk/`m9l?r04$0%_5!(ue]&x7\D/e5bk3^+#1
M!TV!<2cj.$(e4('w0#uwdzg>owxmd"]$cq(@"kl!"ic:fvd..y/%h2kj">
M-w(pl3*/mzi?y/&wseyvdv'o:;_[jskg.v"8;li>0**uc8k;8:h=za*$mnk'
MP@h?f5@]v&'3vxn38;_^30'f?ixc:<['->(je480$)r3!1_)o$'/e%w\s"!
M(($#1?\h,4kz-(7w:3t#q,_+x/)?7ms\;%l)-3,+3lzg\i\mq#>o@b6j'z8m
MBN0M62D['6@m0'ak<)$)t[2sq;wdd=;0xgi?q4,n:nil:anx&<$1f20lsb.g
MZT?K:U@9]D\6V-%EE^'P5#J0CN#ISJS!Z9;[Q@44IM:F:T:[TEH22SU/@>SH^
MH+'OFSLAAJ5&T3XJ[Z85K^<"'@'''''''''@C'0''''L!'`!P'@'`
M'%4#'''3@'''L#'''))0$'''&'''R0,'''LP'@'''@'''#/'P'''"<!
M'''$'''/<#'''(-'$''4'''!P0'@'"'0'\'''';!'')A('%'`
M('!,'&$$$'<p!e('2@!e''0'(!)'$$$'$$D'20''''''''''''''''
M'''''''''''''''$$A04$-,,,4U3'''''''L'2p!+'$$L
M 2p!+'$$L'2p!,'''''''''''''''''''''''''''''''
M''''''''''''''''''''''''''''''''''''''''''''''
M''''''''''''''''''''''''''''''''''''''''''''''
M''''''''''''''''''''''''''!>',<X*'#6'L,/.0@'
M'!$$)''''6@'+'0"+%#8'5'!I'&T'90!S'''3@!E'<'('!2'&\';0!A'&X'
M(('!2'&4'9P!U'&P'80!R''''''''0'"'%'@'"'0'''0'+''''''''
M''''''''''''''''!$@()')'"A'''H0''''('!'(''@'",'P!*9VX@
M''''''''''''''''S?!!X''!S''"'"''''']T*$$$k'00'"p"'"$$$$0
M'-W="PL''p''!'l'w?$$'0#Q\0,!/'Q''(('\?(,,\o$#'@@#q\0("'/'4&q''
Map$""'''@'''?y,0$'''<''%@''@!8'@''p#4\0,"/'q'@$'\51h:7.':7.'
M88!t97?-t\0,!/'q'@('\=0;(p&'0('''!8'm50''!8'@$'''<"''@'^3$W
.''''(P#4\PSS\0,"/'$'
'

end
```

Figure 8.15   The message and attached binary file transmitted using Word Perfect Office to an MCI Mail user. Note that Word Perfect Office automatically UUencoded the binary file

decode the received e-mail, indicating that Word Perfect Office can be used to automatically generate UUencoded files that can be decoded using UUDECODE or a similar program, what would happen if the file transfer resulted in one bit being received in error? That bit error would result in a byte or character being received in error. Then that portion of the line from where the error occurs to the end of the line would be reconstructed erroneously. The effect resulting from this one bit error would depend on the data being transported, and could range from a hard-to-note error if the line was a scan line in an image, to a more pronounced error if the bit error affected characters in a program or wordprocessing document. Thus, one disadvantage associated with the use of some built-in UUencoders, such as the one included in Word Perfect Office, is the absence of a checksum generation capability. This means that the receiving decoder does not have the ability to verify the integrity of the data. In concluding this chapter, we will turn our attention to another built-in UUencoder.

Using WebImage

In Chapter 5 we described the use of several features of the WebImage program from Group42 to manipulate images. One key

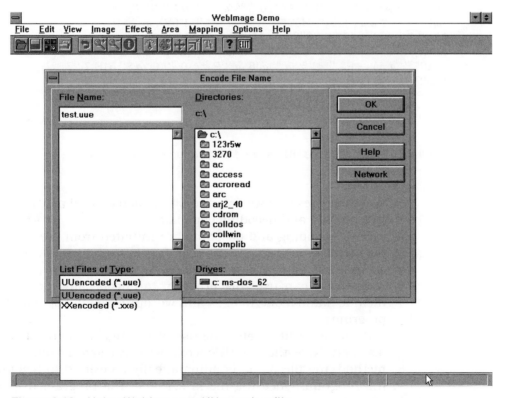

Figure 8.16   Using WebImage to UUencode a file

Encoded by Web Image, Group 42, Inc.
begin 644 TEST.WPD
M_U=0ORH$$$`'!!'"@(!'`'`@4'''#-!'`''('+/!%:WZW\!2VL$$R1^XJ%"='
M%M%M<:O.VX8N8N[>+KQVQ3<^'!+9'^AB[M+%M$_5*]H''GBTS>*G^"5M9'^[[O[I\+^TCFJ[
M@]&Q.,CUCR1D0'WFTZX<\\6Y=8=B2U3[=K=[\J^]B3A/[>:UH_Ld[>:=R;/+*B
M)\65;@\58DMXM_HG9T?HS[-4T6@)ZX1@UYE>>?J'5WKL0R@$$Q6w-M

MHAR.I7U! &PW-35'06N%._/WIG'*0U"^NTK5).S>GW[R8UN!X5'AC0?,BC68N7H
MVVWg;9wZm@(V'LN@)P-JP+Y!_+*x2: 3-L]][G0J8'R#T:S>R)!T[7YZ_0((0
MHDDN>T7109PN8\;-0'QQ9'*^@TEB,#Z=O9ZEH/_M^%>>UUPC'??^PEMMP'+5
M@L7;)!BDl$$<KO=TAOO.XX,B;S)&\R!!(M+>]UD4;F-+EEY*Q2#TIJWK2M9^'<L
M,#)OP]T\C1l!H(xZ&&3,_NDL&@3UB)'6!!Z644Z13*''37'&7&K[T2_N-6Q
M@T@T=399G$$$,$,X@JdH%K!+,64[9]>'D'vNvW8"")V@""Va?=2"4-]5Vo]o&K9)H5$$P3U^
M6@R!!I3!CuQ']PDNvQ:]6=[J]!0$$$$FVD;?$$$$@%w#VO8J6W4/5DCE@@EL;1_)I44'L
M86DD_;-P"'I#.WE18O5::]8""'@"''''''''''@C'0''L!''iP''@''
M'%4#%'''3@''"L#''''))0$$'''&''''R0,''LP''@,''''''Z@'''#/''P'''''<!
M''''$'''' 1<#''''(-'$$'''4''''!P0''@''IO''''\''';''!''''')A('%''
M('!','&&$'<P!E'''('2@'!E''0'(!)'$D'20'''''''''''''''''''''''
M''''''''''''''''''''''$A04$$-,-4U3''''L'2P!++''$$L
M'2P!='$='L'2P!,''''''''''''''''''''''''''''''''''''
M'''
M'''''''''''''''''''''''''''''''''''''
M'''''''''''''''''''''''''''''''''!>'',<X*''#6'L,/.0@''
M'!!')'''''6@'+''0''+'%''%#8'5'!I''&''T'90!S''''''3@''E''<''(!2'&X''
M('.!''2''4''9P!U''&''P'80''R''''''0'''''%''''0''''0'*''''''
M'''''''''''''!$$@'')'''''"A''''H0'''('''!''''@''''''.'P''9@('@''
M'''''''''''''''''''''S?!'X''$$$''''''''']T*$$$$''#''00''P''''''$$0
M'-W=="PL''P''!'''!'L' W?$$$'0#'Q\0,!.'''Q''@('\?(,\$$$#'''@('\?I@Q''
M@P$$$$$$''@=)++P$$$$$'<''%''@''8!'''P'%#4\\0,!'''/''Q''@('\?'@Q''@$$$''
M.''''(P#4#PSz#(''&''''''!8'5''''''''''''''''''''''''''''

Figure 8.17   Viewing the UUencoded file produced by WebImage

set of features that we deferred discussion of until now is its built-in encoding and decoding capability.

Both encoding and decoding are initiated from the program's file menu. WebImage supports both UUencoding and XXencoding as well as their reverse decoding capability. When using Web-Image, you can use this program to encode any binary file to include stand-alone images and wordprocessing documents and programs.

Figure 8.16 illustrates the use of WebImage to encode the file test.wpd. Note that as UUencode was selected as the encoding method, the file name is automatically converted to test.uue by the program. Once the file is selected, the program displays a multi-part format box which allows the user to define a single file or a multi-part encoded file. If a multi-part format is selected, WebImage provides users with the ability to enter the maximum file size for each part, with a default value of 60 000 bytes used by the program. Thus, this program provides a high degree of flexibility in the encoding process.

To determine the compatibility between the WebImage UUencoder and UUDECODE, we will use WebImage to create another UUencoded file. Figure 8.17 illustrates the contents of that file which, with the exception of the header and absence of checksums, produced the same encoded values as the other programs. Thus, UUDECODE can decode the UUencoded file produced by WebImage, and vice versa. Similarly, a file automatically encoded by Word Perfect Office can be restored using the UUdecoder built into WebImage.

As indicated in this chapter, there are several encoding methods you can use to transmit binary files between different e-mail systems as well as to subdivide large files to bypass the e-mail size limitations associated with some electronic mail systems. By carefully selecting and distributing encoding and decoding programs to appropriate e-mail users, you can overcome many of the limitations associated with the transmission of images between subscribers using different electronic mail systems.

SOFTWARE PROGRAM OPERATION

In this concluding chapter we will briefly review the operation and utilization of several freeware, shareware and tradeware image manipulation programs that are contained on the CD-ROM accompanying this book. A freeware program, as its name implies, is made available for public use without requiring compensation. In comparison, a shareware program represents a program distributed on a trial basis for a predefined period of time. This provides users with the ability to try the program and decide whether or not they want to continue the use of the program prior to making a financial obligation by purchasing the program. If they decide that they want to continue to use the program, they normally mail a nominal registration fee, which usually provides them with a printed manual as well as program updates for a predefined period of time, to the vendor. A tradeware program is very similar to a shareware program. One difference between the two is the inclusion of a built-in timer which limits the use of the tradeware program to a predefined number of days or number of executions. To remove this restriction the user telephones, mails, e-mails or faxes a registration fee or credit card number to the vendor and receives a code or key that the user enters into a registration screen to remove the utilization restriction.

If you have frequently accessed different bulletin boards, surfed the World Wide Web, or accessed anonymous FTP sites, you are probably aware that many programs cannot easily be categorized as freeware, shareware or tradeware. Some freeware programs may include a registration reminder display which requires the user to mail a nominal registration fee to the program developer to obtain a more capable version of the program that also removes what (for some persons) is an annoying registration screen. Thus the freeware program, although free, can also be considered as a potential shareware program. In comparison, some tradeware programs are actually shareware programs that expire unless registered. Regardless of the term used, these programs provide a valuable service to the end-user community as they all let you try the program before having to make a decision concerning its use.

In this chapter we will examine the transfer of each program from the CD-ROM to your hard disk.

9.1 THE DISPLAY GRAPHIC UTILITY PROGRAM

The first program selected for inclusion on the CD-ROM accompanying this book is a graphic utility program that can be used to view, convert, manipulate, read and write images and movies. Named DISPLAY, this freeware program was developed by Mr. Jih-Shin Ho.

DISPLAY is a DOS-based program that can also run under the DOS box in a Windows environment, and which supports a very comprehensive series of image file formats. As such, it can be extremely valuable for converting images from one file format to another, as well as to change the parameters associated with a particular image file format. The use of this program requires at least an Intel 386-based PC, with a 486 with 8M RAM highly recommended.

Program storage

The DISPLAY program is stored on two files on the CD-ROM. The first file, DISP189A.ZIP, represents the compressed archive that contains the basic program, documentation, configuration information and drivers for a large number of video cards. The second file, DISP189B.ZIP, contains a large number of fonts that are usually used to create contact sheets. If you are using DISPLAY primarily to convert images from one image format to another, you will not require the use of the second archive but can consider leaving it on the CD-ROM. Otherwise, its expansion will result in the generation of 263 files that will use approximately 500 Kbytes of disk storage.

Operation

As the best way to obtain an overview of the capability of a program is from its use, let us do so. Change to DOS prompt and navigate to the DISPLAY directory on the CD-ROM enter the command DISPLAY to execute the program.

The initial execution of DISPLAY results in a screen display that shows the contents of the default directory. Figure 9.1 illustrates the initial screen display on the author's PC. Note that you can move the highlighted bar over a particular file to select a file.

The DISPLAY program is a command entry program requiring the pressing of certain keys and key combinations to perform predefined operations. Table 9.1 lists the major key and key combination operators supported by DISPLAY. As we discuss the operation of the program we will also refer to the use of a few additional keys that are not included in Table 9.1.

Assuming that we have a group of images on a diskette, we would press the F2 key to change drives. Figure 9.2 illustrates the resulting display on the author's PC which is connected to a LAN. Thus, in addition to local drives A through D being available for selection, the program also displays network drives F and S–Z. By moving the cursor over the appropriate drive letter and pressing the Return key, the program will display the files on the selected drive.

[Auto Read]		[Manually Read]	[Quit]	[Re-Show]	[Write]
..		change	35501	char23bi.fnt	9704
acumos.asm	13857	char11.fnt	2968	char23i.fnt	9290
acumos.grn	547	char11b.fnt	2990	char30.fnt	13524
aheada.asm	4253	char11bi.fnt	3001	char30b.fnt	15054
aheada.grd	287	char11i.fnt	2968	char30bi.fnt	15144
aheadb.asm	4055	char14.fnt	3934	char30i.fnt	14064
aheadb.grd	267	char14b.fnt	4060	char40.fnt	24784
ati.asm	7493	char14bi.fnt	4004	char40b.fnt	26464
ati.grd	375	char14i.fnt	3920	char40bi.fnt	27144
		char16.fnt	4808	char40i.fnt	24864
ati_16md.asm	16045	char16b.fnt	5064	chips.asm	4243
ati_16md.grn	979	char16bi.fnt	5096	chips.grd	282
atigupro.asm	11534	char16i.fnt	4920	cirrus54.asm	15717
atigupro.grn	513	char18.fnt	5706	cirrus54.grn	700
atiultra.asm	13068	char18b.fnt	6858	c15426.asm	11781
atiultra.grn	1208	char18bi.fnt	6804	c15426.grn	549
ativga.asm	11803	char18i.fnt	5850	config.dis	29896
ativga.grn	565	char23.fnt	9244	copying.cb	4360
		char23b.fnt	9612	copying.dj	1625
8 bits, 640 x 400, 325 file, 0 tag(0 bytes)					

Figure 9.1 The initial execution of DISPLAY displays the files in the default directory

Table 9.1   Major DISPLAY key and key combination operators

F1	:	Show DISPLAY.DOC	ALT-A :	Re-read directory. Keep desc.
Ctrl-F1	:	Show user-defined help	ALT-B :	Re-read directory. Discard desc.
F2	:	Change disk drive	ALT-C :	Copy single file (ignore tags)
F3	:	Change filename masks	ALT-D :	Delete single file (ignores tags)
F4	:	Change parameters	ALT-E :	Edit description entry
F5	:	Special effects	ALT-M :	Move single file (ignores tags)
F6,m,M	:	Move file(s)	ALT-S :	Save config file
F7	:	Make Directory	ALT-T :	Tag all files
F8,d,D	:	Delete file(s)	ALT-U :	Untag all files
F9	:	Open menubar	ALT-W :	Write description entries
ESC,F10	:	Quit	ALT-X :	Quit program without prompting
			ALT-Z :	Shell to DOS

CTRL-A -- CTRL Z	: Change disk drive.
0 .. 9; SHIFT-A .. Z	: Jump to first file whose name starts with this character.
TAB	: In mode selection screen : Change process target.
	In file selection : Show previously loaded image.
Ins	: change display type (8, 15, 16, 24 bits).
PageUp/Down	: Move one page.
Left/Right arrow	: change display type In 'screen' & 'effects' menu.
BACKSPACE	: Go up one level in the directory tree.
+	: Group tag
−	: Group untag

Figure 9.3 illustrates the display of files on drive A, with the highlighted bar shown over the file Fig1-2.TIF. You can use any arrow key to move the highlighted bar over a different file entry as well as the Home and End keys to rapidly position the bar over the left-most or right-most entry. Once you have positioned the highlighted bar over an appropriate image file entry and pressed the Return key, the program will display a screen of information consisting of the current resolution mode of the image as well as other resolutions that the image can be converted to. Figure 9.4 illustrates the image resolution screen display.

In examining Figure 9.4 note that the display informs you of the three key keys that you can use: RETurn, SPaCe and ESCape. Press-ing the Return key results in the display of the image in the selected resolution mode. Pressing the Escape key returns the display of the prior screen, and pressing the Space key results in the program, providing you with the ability to save the image in a different format.

Figure 9.5 illustrates a portion of the Image Write capability of the DISPLAY program. This screen was generated as a result

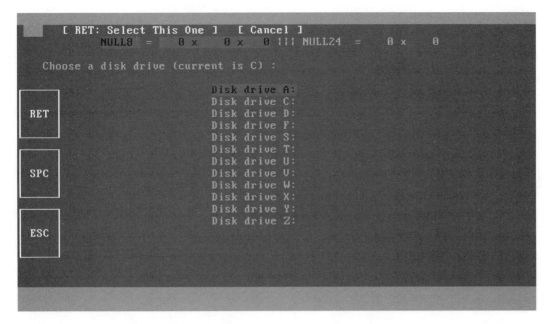

Figure 9.2 Through the use of the F2 key, you can select a different disk drive

```
[ Auto Read ]    [ Manually Read ]    [ Quit ]    [ Re-Show ]    [ Write ]
fig1-2.tif         64292   fig4-9.tif        48954
fig3-1.tif         31844
fig3-10.tif        16211
fig3-11.tif        16944
fig3-2.tif         21763
fig3-3.tif         29625
fig3-4.tif         19640
fig3-5.tif         28063
fig3-8.tif         83268
fig3-9.tif         33702
fig4-1.tif         33411
fig4-10.tif        57735
fig4-2.tif         24953
fig4-3.tif         18440
fig4-4.tif         55072
fig4-5.tif         54572
fig4-6.tif         94531
fig4-7.tif         55848
fig4-8.tif         33108
         8 bits,  640 x  400, 20 file, 0 tag(0 bytes)
```

Figure 9.3 Once a drive or directory has been selected, moving the highlighted bar over a file name and pressing the Return key selects the file

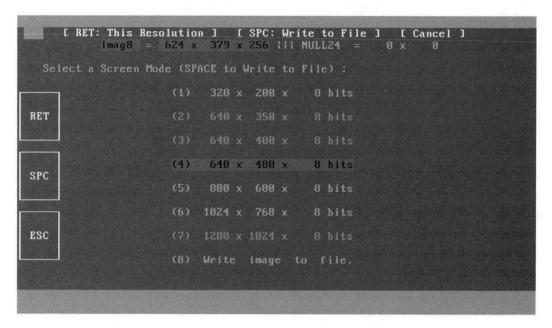

[RET: This Resolution]   [SPC: Write to File]   [Cancel]
        Imag8 = 624 x 379 x 256 !!! NULL24 =   0 x   0

Select a Screen Mode (SPACE to Write to File) :

                    (1)   320 x 200 x   8 bits

RET   (2)   640 x 350 x   8 bits

                    (3)   640 x 400 x   8 bits

                    (4)   640 x 480 x   8 bits

SPC   (5)   800 x 600 x   8 bits

                    (6)   1024 x 768 x   8 bits

ESC   (7)   1280 x 1024 x   8 bits

                    (8)   Write image to file.

Figure 9.4   The image resolution screen denotes the current resolution mode of the selected file as well as other resolutions supported

of pressing the Space bar after a prior image was selected including a display mode for the image. Although 16 image formats are shown in Figure 9.5, DISPLAY provides support for additional image formats, with those formats becoming visible as you scroll the highlighted bar downward. Assuming that you want to convert the previously retrieved TIF image as a GIF image, you would press the Return key once the highlighted bar is over the first entry in Figure 9.5.

One of the key features of DISPLAY is its support for an extremely large number of options for most of the image file formats it supports. This is illustrated in Figure 9.6, which shows the screen display generated by selecting the CompuServe GIF entry previously shown in Figure 9.5. Note that the highlighted bar over 'Use interlace' permits you to change the default setting of No to Yes to create an interlaced GIF image. Although GIF uses a lossless compression method, DISPLAY allows you to alter several characteristics of the image to generate a reduced and lossy image. Although doing so can significantly reduce the resulting size of the file, this author recommends that you may be better served converting to JPEG, unless your application requires a GIF file format.

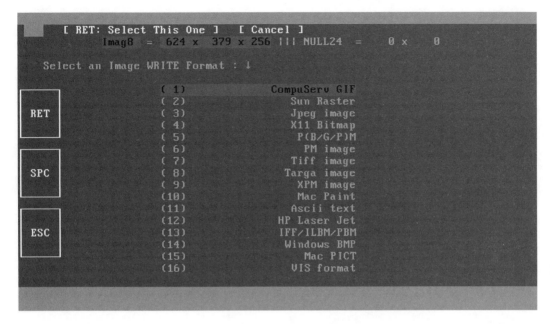

Figure 9.5 Once a resolution mode has been selected, pressing the Space bar generates the program's Write display. From this display you can select an appropriate image format

Although this brief description of DISPLAY has only scratched the surface of the capability of the program, it provides you with an overview of its use to convert images from one format to another. As different users will have different image conversion requirements, the primary purpose of this section was to introduce you to the image conversion capability of the program by converting a previously stored TIF file into a GIF file. You can also use this comprehensive program to prepare a slide show of images, as well as to view the resulting show, crop images, rename files, and perform a variety of other operations that you can determine by reading the documentation files included in the program archive.

9.2 WEBIMAGE AND I SPY PROGRAMS

In Chapter 5 we used a Windows 3.1 version of the WebImage program from Group 42 Inc. to manipulate several images as well creating a thumbnail catalog of images. In Chapter 8 we returned to the use of the Windows 3.1 version of WebImage to UUencode a binary file. Thus, prior to this chapter we obtained an appreciation for the capabilities of the Windows 3.1 version of WebImage.

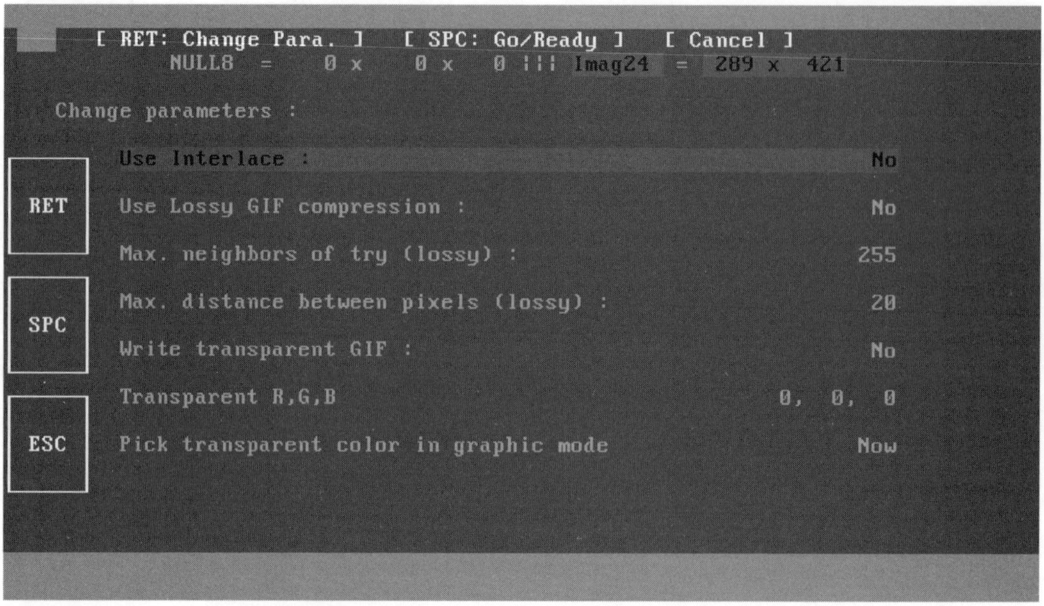

```
[ RET: Change Para. ]    [ SPC: Go/Ready ]    [ Cancel ]
        NULL8  =    0 x    0 x   0 ||| Imag24  =   289 x  421

  Change parameters :

       Use Interlace :                                      No

RET    Use Lossy GIF compression :                          No

       Max. neighbors of try (lossy) :                      255

       Max. distance between pixels (lossy) :               20
SPC
       Write transparent GIF :                              No

       Transparent R,G,B                               0,  0,  0

ESC    Pick transparent color in graphic mode              Now
```

Figure 9.6 DISPLAY supports a number of options for most file formats it creates. When creating a GIF file, you can create the file as an interlaced image as well as selecting other options indicated on this screen

When Group 42 created the Windows 95 version of WebImage, the thumbnail catalog features were moved out of that program and incorporated into their I Spy product. Thus, the CD-ROM containing the Windows 95 version of WebImage also includes the I Spy program. In addition, for persons using Windows 3.1, the CD-ROM also contains the Windows 3.1 version of the original WebImage program. However, readers should note that thumbnail catalogs made using the Windows 3.1 version of WebImage (WebImage v1.72) are not compatible with the catalog system used by I Spy.

Trial period

Both the Windows 3.1 and Windows 95 versions of WebImage and the Windows 95 version of I Spy can be considered as tradeware software. Each product has a 10-day evaluation period, after which the saving, caching, and catalog functionality is disabled until the product is registered. If you register WebImage v1.72, you can automatically upgrade to the Windows 95 version of the program to include I Spy for free. Readers can obtain a registration key from Group 42 via telephone, e-mail, World Wide Web, or fax via the following:

Toll Free:	1-800-520-0042 (U.S. only)
Telephone:	1-513-831-3400
e-mail:	info@group42.com
WWW:	www.group42.com
Fax:	1-513-831-8289

Program installation and operation

In this section we will focus our attention on the installation and operation of both Windows versions of WebImage. First, we will examine the installation of the Windows 95 version of the program and briefly examine its operation because it is very similar to the Windows 3.1 version which we used in Chapters 6 and 8. Once this has been accomplished, we will examine the installation of the Windows 3.1 version of the program.

Windows 95

Two files are stored on the CD-ROM that represent the Windows 95 version of WebImage. Directory **Web201** on the CD. The first file is INSTALL.EXE, and the second file is INSTALL.A01. INSTALL.EXE is equivalent to disk 1 of the two disk program, with the file INSTALL.A01 representing the contents of the second disk.

Under Windows 95 you would select Run from the START menu. Then, if the CD-ROM is installed in drive D on your computer, you enter D:INSTALL and press ENTER to install this program.

Figure 9.7 illustrates the initial Windows 95 Welcome! display screen resulting from the selection of the INSTALL.EXE file in the START menu. In examining Figure 9.7 note that the installation process will install both WebImage and I Spy on your computer. In addition, note that the installation program recommends that you should exit any previously activated programs before continuing with the installation. This screen tells you to press the Ctrl + Esc keys to switch to any previously opened programs and close those programs after saving your work before continuing

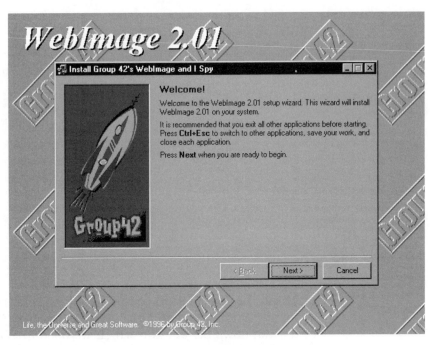

Figure 9.7 The Group 42 Install window recommends that you close any previously opened applications before installing WebImage and I Spy

with the installation. Once this has been accomplished you would return to the Group 42 install window shown in Figure 9.7 and click on the button labeled Next to continue the installation.

Once you close any previously opened applications and click on the button labeled Next, the program's license agreement will be displayed. A portion of this agreement is shown in Figure 9.8. If you accept the terms of the agreement, you would click on the button labeled Next, otherwise you would click on the button labeled Cancel to exit the installation process. Assuming that you agree to the license agreement terms the installation program will display a new window labeled Program icon folder. This window is illustrated in Figure 9.9.

As illustrated in Figure 9.9, the WebImage installation program uses the default Group 42 for the name of the folder it will create and transfer its files into. Although Group 42 is the name of the developer of WebImage and I Spy, you may wish to consider a more meaningful folder name, such as WebImage, Graphics, Images, or a similar folder name that may be easier to remember if you only periodically intend to use the program and have a large number of existing folders. To use a different folder name you can either type a new name into the name box or click on the Browse

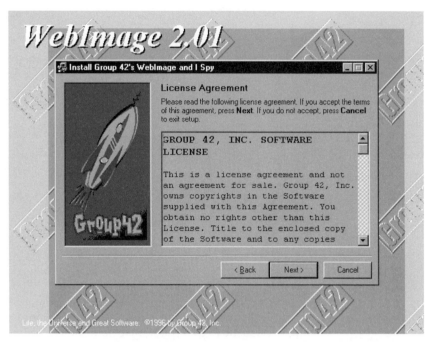

Figure 9.8 You can scroll through the license agreement and click on the button labeled Next if you accept its terms. Otherwise, clicking on the button labeled Cancel terminates the installation process

button to select a previously created folder to use. After you have selected the folder name, you would again click on the button labeled Next to continue the installation process. As files are extracted from the installation program and copied to your hard drive, the program will display a progress screen similar to the one shown in Figure 9.10.

Once appropriate files for WebImage and I Spy have been copied to your hard drive, the program will display another dialog box. This box provides you with the option to use Web Image as the default application for viewing images on your computer and add Group 42's home page to Microsoft's Internet Explorer if that browser is stored on your computer. This box includes two check marks set to perform those operations as illustrated in Figure 9.11. If you do not want one or both functions performed, you can click on either or both check boxes to remove the checks. As this is the last dialog box displayed during the program's installation process, the buttons at the bottom of the display include one labeled Finish instead of Next.

If you agree with the first option listed in Figure 9.11, the program will change the icons used to represent the icons used to

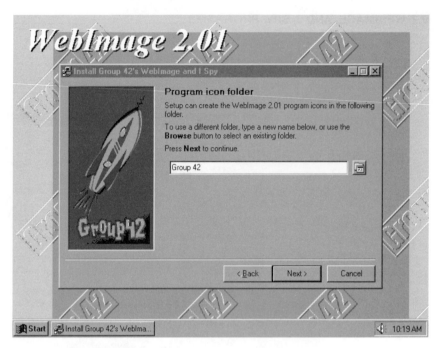

Figure 9.9 Through the use of the window labeled Program icon folder you can select the default Group 42 folder name, enter a new folder name, or use the Browser button to select a previously created file folder

Figure 9.10 The WebImage progress screen indicates the status of files being copied during the program installation process

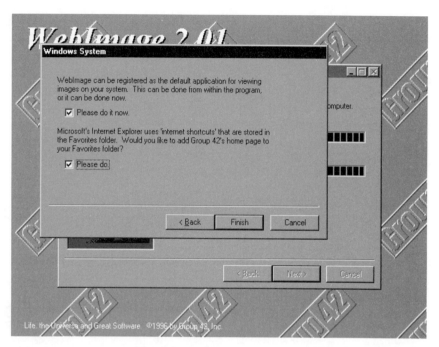

Figure 9.11 You can register WebImage as your default application for viewing images on your computer and add the developer's home page to Microsoft's Internet Explorer

represent images that it supports that were previously created on your computer. For example, if you had used Collage Image Manager, that program created square icons that were labeled with the name of image files used by the program. WebImage will change each of those icons into a shape that resembles a pair of glasses with what appears to this author to be a spider dangling between the frames. Whether or not you like this icon will depend on your personal preferences, and you may wish to consider unchecking the default check mark at the top of Figure 9.11 if you prefer to leave your image icons as is on your computer.

After the program installation process is complete, a folder labeled Group 42 will be displayed, assuming that you agreed to use Group 42 as the default folder label. Figure 9.12 illustrates the Group 42 folder. In examining the contents of the Group 42 folder illustrated in Figure 9.12 you will note that eleven objects are in the folder. The icon labeled WebImage that looks like a pair of glasses with a spider dangling in the middle represents the program, and the icon is the one that the installation program will use to replace previously created image icons if you agreed to

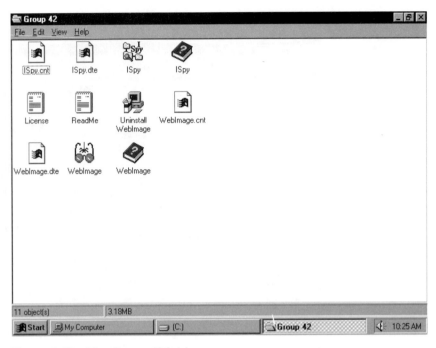

Figure 9.12 The Group 42 folder

having the program registered as the default application to view images on your computer.

The icon labeled ISpy represents the I Spy program that you can use to create thumbnail catalogs. If, after using WebImage during its 10-day trial period, you decide that you do not want to register the program, you can gracefully remove it, I Spy and the Group 42 folder by clicking on the icon labeled Uninstall WebImage. The icons labeled License and ReadMe, as you might expect, contain information about the program license and registration data, whereas the two icons that appear as books with question marks represent help files that can be accessed directly or through WebImage or I Spy.

The Windows 95 version of WebImage is very similar to the Windows 3.1 version in that it is very intuitive to use and can be mastered quickly. Thus, in this section we will take a quick look at the Windows 95 version of WebImage, including opening an image file stored on disk and saving the file using a different image file format.

To open a file you would select the Open option from the program's File menu, similar to the manner by which a file is opened when the Windows 3.1 version of the program is used. The

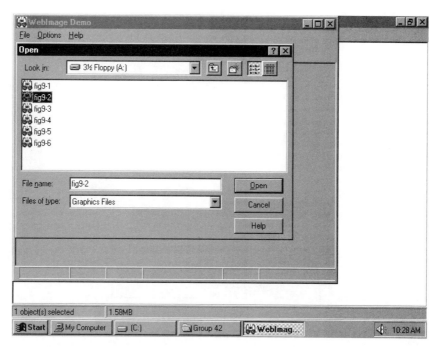

Figure 9.13   Using the Open option from the file menu to select an image

key difference between the use of most Windows 95 and Windows 3.1 program options is the manner in which each version of Windows displays information within a dialog box. Figure 9.13 illustrates the selection of the Open option from the program's File menu. In this example drive A was selected as the default drive, resulting in the program displaying the images stored on the disk in drive A. Note that under Windows 95 Look in: is used as a replacement for the label Drives used in Windows 3.1. Thus, most differences between the Windows 95 and Windows 3.1 versions of the program are not significant.

In Figure 9.13 the file Fig9-2 is shown as being selected, resulting in its name placed in the box labeled File name. Assuming that you want to save the file using a specific file format, you would either select Save or Save As from the file menu, with the former option resulting in the image being stored under its existing file name and native format. If you want to vary the existing format, you would use the Save As option.

Figure 9.14 illustrates the use of the Save As option with the Save As type drop-down menu displayed and the JPG option selected in that display. Similar to the Windows 3.1 version of this program, clicking on the button labeled Options after JPG has been selected would allow you to specify a Quality factor. In

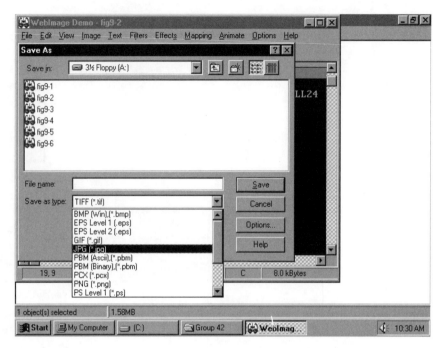

Figure 9.14   Viewing the image file formats supported by the Windows 95 version of WebImage

addition, WebImage also supports the creation of progressive JPG images, so you can consider their use as a replacement for interlaced GIF if you desire. As the use of WebImage is intuitive and a comprehensive manual is available once you have registered this program, it is suggested that readers try this program as this is the best way to become familiar with its use.

Windows 3.1

As previously mentioned, the version of WebImage that operates under Windows 3.1 includes a built-in thumbnail catalog capability. Thus, this program version does not include a separate I Spy program.

The program version of WebImage developed for use under Windows 3.1 can be found in the directory **Web172** on the CD-ROM. You can install the contents of this self-extracting file by first selecting RUN from the File menu and entering D:WI16_172.EXE in the command line of the dialog box labeled RUN, again assuming that the CD-ROM is installed in drive D on your computer. Pressing the OK button after the previously described actions results in the display of the WebImage 16-bit installation screen as illustrated in Figure 9.15. Note that this display allows you to select the default directory WebImage or to enter a different directory for the storage of the program and its associated files. In addition, note that the display informs you that this version of WebImage requires almost 2 Mbytes of storage. When you click on the button labeled Continue, the installation program will extract the files contained in the self-extracting file and display the progress of the extraction effort. After the file extraction process has been completed, the program will generate a set of icons that will be added to the Program Manager group that has the default name Group 42. Figure 9.16 illustrates this display. Note that you can enter a different name for the group or click on the button labeled Create to accept the default. As discussed earlier in this section, when we covered the installation of the Windows 95 version of this program, if you only periodically use this program and have a large number of

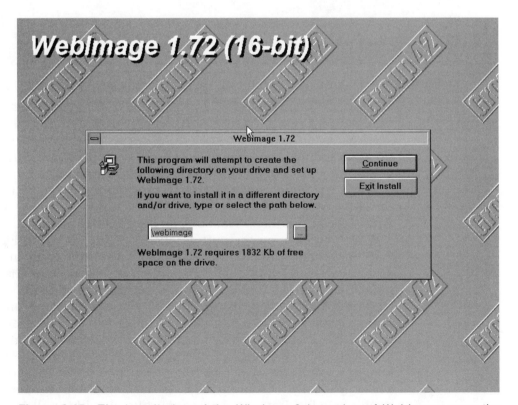

Figure 9.15 The installation of the Windows 3.1 version of WebImage uses the default directory WebImage to store the program and its associated files

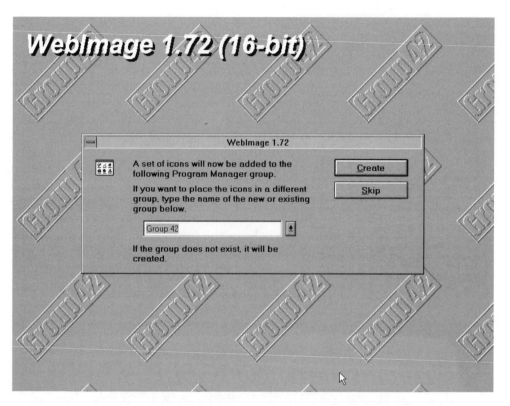

Figure 9.16 You can accept the default Group 42 for the icon group added to the Windows 3.1 Program Manager group or enter another group name

program groups, you may wish to consider replacing the default group name with one easier to recognize as an image-manipulation activity. Thus, you may wish to consider replacing the use of the developer's name for the group by the program name WebImage or another term, such as Images or Graphics.

When you click on the button labeled Create shown in Figure 9.16, the installation program will rapidly create the group and its associated icons and then cover that screen with a display indicating various options for registering the program and obtaining technical assistance. After you have clicked on the button labeled OK, you can observe the newly created group of icons. This group, which is illustrated in Figure 9.17, contains four icons. The icon labeled WebImage.exe represents the Web-Image program that we used in Chapters 6 and 8. The second icon whose partially obscured label is WebImage.hlp is the program's help file that can be accessed either directly by clicking on that icon or indirectly through the program. As you may surmise,

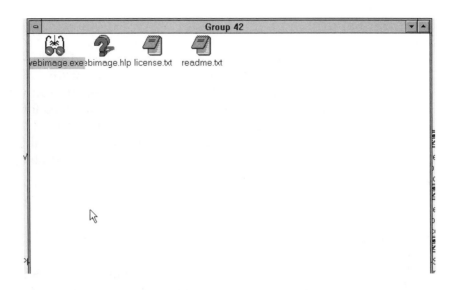

Figure 9.17 The Group 42 program group contains four icons to include the main program, its help file and two text files

the two notepad-shaped icons with the file extension .txt are text files that when double-clicked invoke the Windows 3.1 notepad to display their contents. As the use of the Windows 3.1 version of WebImage was previously covered in Chapters 6 and 8, we will not review its use in this section.

9.3 PAINT SHOP PRO

Through the courtesy of JASC Inc. two versions of Paint Shop Pro are included on the CD-ROM accompanying this book. The first version, which can be found in the directory **psp311**, represents the 16-bit version of this powerful image viewing, editing and image conversion program. This 16-bit version operates under Windows 3.xx or Windows NT 3.51. The second version of the program which can be found in the directory **psp41** represents the 32-bit version of Paint Shop Pro 4.10 for use with Windows 95 or Windows NT 4.0.

The developer of Paint Shop Pro recommends the use of a 386 or faster CPU with at least 4 MB of RAM and a 16-color display capability to operate the 16-bit version of the program. The minimum system requirement for Paint Shop Pro 4.10 is a 486 or faster computer with at least 8 MB of RAM and a 256-color display capability.

Shareware information

Both versions of Paint Shop Pro included on the CD-ROM that accompany this book represent shareware. As previously mentioned in this book, shareware represents software that you can try before you buy. Paint Shop Pro is provided for you to try out for 30 days free of charge. If you continue to use either program provided on the CD-ROM after a 30-day trial period you must register the program by following the instructions included with each program.

Installation

The installation of either version of Paint Shop Pro from the CD-ROM involves similar steps. Thus, we will primarily focus our

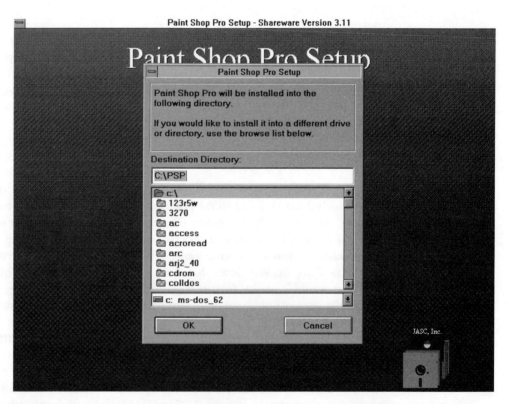

Figure 9.18   The Paint Shop Pro Setup program uses the default directory PSP to store program files

attention on the installation of the Windows 3.11 and Windows NT 3.51 version of the program. Later in this section we will briefly note the slight changes necessary to install the Windows 95 and Windows NT 4.0 version of the program.

The first step in the installation process is to explode the zipped archive. Assuming that your CD-ROM drive is drive D and that a copy of PKUNZIP is located on drive C under the directory PKZIP, you would enter the following sequence of DOS commands: Start Windows and select the Run option from the Program Manager File menu. You would then select the file SETUP.EXE from the directory PSP and click on the OK button.

The Paint Shop Pro setup program will display an initial dialog box similar to the one shown in Figure 9.18. From that dialog box you can define the destination directory for the installation of the program or accept the default of PSP. Assuming that you accept the default and click on the button labeled OK, the program will display a progress or status screen indicating the status of the program's setup effort. Within approximately 30 seconds, the program will complete its initial setup effort and prompt you concerning the addition of appropriate icons for Paint Shop Pro to the Windows Program Manager, as well as allowing you to specify the name of the Program Manager group where the icons will be placed. The setup program uses the default Paint Shop Pro for the group name; however, you can easily change that name if you so desire.

Figure 9.19 illustrates the icons created by the Paint Shop Pro setup program. The highlighted icon, labeled Paint Shop Pro 3, represents the main program. The icon labeled PSP Browser provides access to a utility program that can be used to browse selected directories and which rapidly creates thumbnails of images stored in those directories. The two icons with the suffix label Readme are, as the suffix denotes, text files which when clicked on display program information. The fifth icon, which has the label prefix Uninstall, will gracefully remove the program and its associated files when double-clicked on.

Figure 9.19   Paint Shop Pro program icons

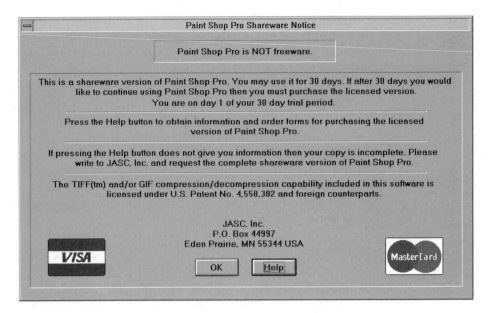

Figure 9.20 Paint Shp Pro shareware notice

Program operation

Although the use of a few features from this feature-rich program was discussed earlier in this book, we will take a new look at Paint Shop Pro to illustrate both its ease of use as well as some of its more unusual features.

The operation of Paint Shop Pro is accomplished by double-clicking on the icon labeled Paint Shop Pro 3 shown highlighted in the right portion of Figure 9.19. As a reminder to users that have not registered the program, Paint Shop Pro displays a shareware notice similar to the one illustrated in Figure 9.20. You can obtain a list of vendors located throughout the world, from which you can purchase Paint Shop Pro by selecting the Purchasing option from the program's Help menu. You can use either Visa or MasterCard to purchase a product directly from JASC Inc. in U.S. funds or select a vendor outside the U.S. for payment in an appropriate currency other than dollars.

Working with images

When you click on the button labeled OK on the Shareware Notice screen, the program displays a menu bar across the top of the screen with certain options highlighted, representing those then

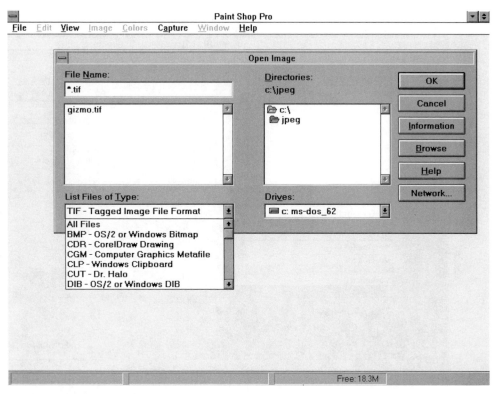

Figure 9.21 The Open Image dialog box allows you to retrieve images stored in a variety of formats

available for selection. To work with an image you must have one; thus, you would use the Open option from the file menu to select an appropriate image.

Figure 9.21 illustrates the Open Image dialog box after the Open option has been selected from the File menu. In Figure 9.21 the List Files of Type menu was pulled down to illustrate how you can scroll through the file types supported by the program. Assuming that the author's dog is still friendly after its image abuse in earlier chapters, let us select poor old Gizmo again as it is the only TIF file in the selected directory.

Assuming that we have retrieved the previously mentioned image file, we can convert it into a different file format by using the program's Save As option. That option is also selected from the program's File menu.

Figure 9.22 illustrates the use of the Paint Shop Pro Save As option to save a previously retrieved TIF file as a Word Perfect Graphic (WPG) Version 5.1 file. In examining Figures 9.21 and 9.22, note that Paint Shop Pro is network-aware, which enables

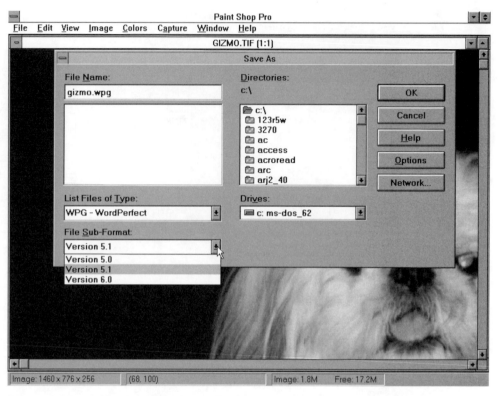

Figure 9.22 Using the Save As option from the File menu to create a new image file format

the program to support files stored on a network or to save files to a network device. Now that we have an overview of the ease in which we can manipulate file formats via the use of Paint Shop Pro, let us turn our attention to a most interesting feature of the program known as the Deformation Browser.

The Deformation Browser

Built into Paint Shop Pro is a feature referred to as the Deformation Browser. Through the use of the Deformation Browser you can examine the effect of potentially applying different predefined shapes to an image. If you like the effect of the application of a shape you can then apply the effect to the image. The Deformation Browser is selected from the Image menu in the program's menu bar.

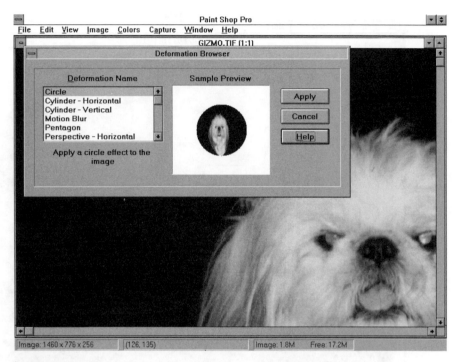

Figure 9.23 Viewing the effect of a circle deformation

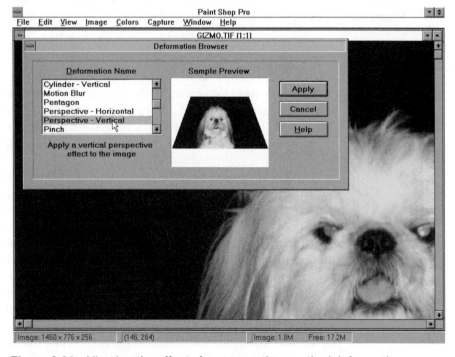

Figure 9.24 Viewing the effect of a perspective-vertical deformation

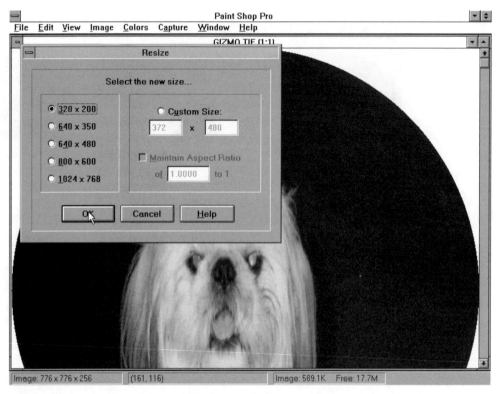

Figure 9.25 Resizing the circle deformation

To illustrate the use of the built-in Deformation Browser, let us examine its use. Figure 9.23 illustrates the preview obtained by selecting the circle as the deformation, and Figure 9.24 illustrates the use of the Perspective-Vertical deformation. Needless to say, you can create some rather interesting images through the use of the program's Deformation facility.

By examining the sample preview as you alter the selection of deformation names, you can determine if you want to apply a particular effect to a previously selected image. Assuming that you find an effect that fits your requirements you would click on the button labeled Apply. If the results are acceptable but the image is too large or too small, you can select the Resize option from the Image menu to alter the size of the image.

Figure 9.25 illustrates the resizing of the previous circle deformation TIF image to a 320 × 200 size. Figure 9.26 illustrates the resulting image, making it appear that good old Gizmo is about to pop out of the circle on the screen and onto your lap. Now just imagine placing this or a similarly altered image on your Web homepage!

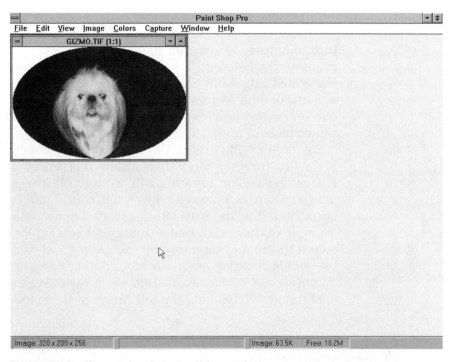

Figure 9.26 The resized circle deformation

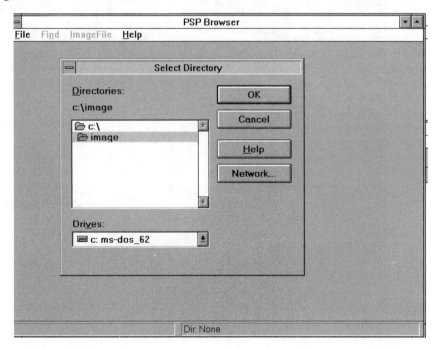

Figure 9.27 From the file menu the Select Directory option is used to select the directory from which a catalog of thumbnails will be created

The Browser program

In concluding this examination of Paint Shop Pro we will briefly examine the Browser. The Browser is a separate program that is integrated into Paint Shop Pro whenever you request the display of a thumbnail. As previously mentioned, the Browser program is used to rapidly preview images in a selected directory as it creates thumbnails of image files in a selected directory.

Figure 9.27 illustrates the initial screen display of the Paint Shop Pro Browser after this author had selected the directory labeled Image for browsing. By simply clicking on the button labeled OK the browser will create thumbnails of image files encountered in the selected directory. Figure 9.28 illustrates an example of the creation of a thumbnail catalog for eleven images stored in the root directory on the author's hard drive. Note that by double-clicking on a thumbnail the Browser program will automatically invoke Paint Shop Pro, resulting in that program opening and displaying the full image of the selected thumbnail.

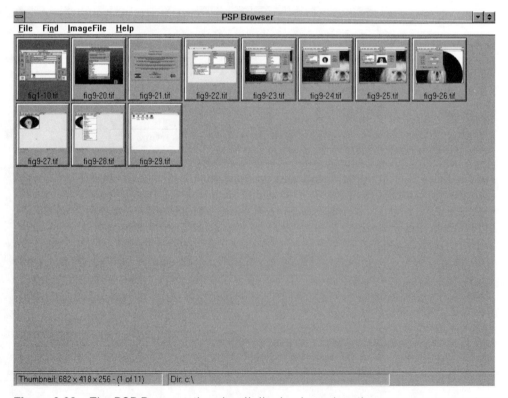

Figure 9.28 The PSP Browser thumbnail display based on the program encountering eleven image files in the root directory on the author's hard drive

Although Paint Shop Pro and its PSP Browser are chock-full of other valuable features, space constraints force this author to suggest that readers use the programs to become aware of those features. When you do so, you should remember to register the program if you intend to use it after its 30 day trial period.

Version 4.10

In concluding this discussion of Paint Shop Pro we will briefly discuss the installation and operation of the 32-bit version of the program developed for use under Windows 95 and Windows NT 4.0. To install this version, run the setup.exe file which can be found in the **psp41** directory on the CD-ROM. In addition, under Windows 95 or Windows NT 4.0 you would select the Start button and then select the Run entry before typing in the command C:\PSP\SETUP.

Version 4.10 includes several new features, some of which may justify its use instead of the 16-bit program. For example, Version 4.10 adds support for several new file formats, including Macintosh PICT, Progressive JPEG and PNG, both with and without transparency. In addition, Version 4.10 adds a variety of special effects, such as drop shadow, chiseling, hot wax coating, cutout and seamless tiling which, when applied to images, may give your documents or Web server the image you require.

9.4 UUENCODE/UUDECODE

As noted earlier in this book, UUencoding represents a method to translate the contents of binary files into printable plain-text characters that can be attached to e-mail messages. The basic character encoding process used by a UUencoder is to break groups of three eight-bit characters into four six-bit characters and then add 32 to each six-bit character. This encoding technique ensures that a binary file is mapped into a series of printable ASCII characters that more than likely can flow through most e-mail gateways. As some gateways in their quest for efficiency remove trailing spaces, most UUencoders also convert spaces into the back-quote character. Otherwise, the removal of one or more trailing spaces could play havoc when the UUdecoding process is performed.

The UUencoding and UUdecoding program pair discussed in this section and included on the CD-ROM accompanying this book were developed by Richard Marks of Bryn Mawr, PA. The freeware version of the program pair includes support for UUencoding and UUdecoding as well as XXencoding and XXdecoding. A more complete version of the program is available from Mr. Marks for a nominal registration fee and adds support for BASE 64 encoding and decoding.

Program installation

The file UUENCODE.EXE stored on the CD-ROM that accompanies this book is a self-extracting file. Assuming that the CD-ROM is located on drive D, you would enter the following DOS commands to create the directory UU, change the default drive to that directory, and execute the self-extracting file so that its contents are extracted into the subdirectory UU on your hard drive:

```
MD\UU
CD\UU
D:UUENCODE C:\UU
```

The execution of the self-extracting program will result in the creation of seven files that will be placed in the directory UU on your hard drive. The files UUENCODE.EXE and UUDECODE.EXE represent UUencoding and UUdecoding programs. The file UUTECH.TXT provides a description of the more technical aspects of UUencoding and UUdecoding, whereas the file UUSER.TXT provides documentation for the use of the UUEN-CODE and UUDECODE programs. Concerning the latter, it is highly recommended that persons that need to use one or more of the many optional features included in the programs should print that file. Doing so will provide you with a comprehensive reference to the use of each program including their options. The files UUWIN31.REG and UUWIN95.REG provide information for the setup of the program under those operating systems, whereas the file UUSETUP.EXE allows for program setup under DOS and other operating environments. As the program documentation covers the use of the program options and we previously investigated its use in an earlier chapter, we will conclude this section by examining the use of the UUENCODE/UUDECODE program pair with a technique that may reduce the cost of using an e-mail system to transmit binary files if the mail system is similar to MCI Mail and bills for e-mails based on the number of characters in a message.

Using UUENCODE/UUDECODE

In using the UUENCODE/UUDECODE program pair we will return to the use of the WordPerfect document containing the message 'This is a test' with each character stored in a bold 32-font size. As we noted earlier, the file TEST.WPD required 1229 bytes of storage and the UUencoded file TEST.UUE required 1918 bytes of storage. Rather than repeat the creation of a previously created UUencoded file, let us see if we can send it via an e-mail attachment which, if we are being billed by file size, can save us some money. To do so, let us first compress the original binary file. Although we will use PKZIP in the following example, you can also use another archiving and compression performing program if you wish. Using PKZIP we would enter the following command:

PKZIP TEST.ZIP TEST.WPG

The preceding command line entry would create the zip file TEST.ZIP containing a compressed copy of the TEST.WPG file. The result of this action is a file (TEST.ZIP) that requires 1024 bytes of storage in comparison to the original binary file that required 1229 byes of storage. Now that we have reduced the size of the file by approximately 26%, let us UUEncode it. To do so we would enter the following command:

UUENCODE TEST.ZIP TEST.UUE

The newly UUencoded file TEST.UUE will require 1629 bytes of storage, in comparison with the 1918 bytes of storage required when the original Word Perfect file was UUencoded. Thus, compressing the file before UUencoding saves 289 bytes. Although this does not appear to be significant, let us consider a more practical application for compressing before encoding, which this author recently experienced. When developing a presentation using Microsoft's PowerPoint program, several images were scanned and added to the presentation, resulting in a .PPT file (.PPT is the extension used by Power Point) requiring 15 524 864 bytes of storage. The compression of that file reduced its size by approximately 92% to 1 290 803 bytes. Although still a large file, its UUencoding was much more practical than if it was performed on the original file. In addition, if an e-mail system that bills for messages based on the length of the message were used, the resulting cost to transmit the PowerPoint presentation would be significantly less than if the non-compressed file were UUencoded and e-mailed.

Two additional considerations that may justify compression prior to UUencoding are the operation of e-mail gateways and the effect of large files on those gateways. First, many e-mail gateways require 1.5 to 2 bytes of RAM per byte in an e-mail file, because the gateway stores the file before converting and transmitting it onto a new e-mail system. This means that the transfer of very large files, if permitted, may require a costly RAM upgrade to the e-mail gateway. Secondly, many gateways operate on a first-in, first-out basis. This means that the transfer of a large file via a relatively low-speed communications facility could result in significant delays to the transfer of messages following a message with a large file attached to it. Thus, compression and UUencoding provide the ability to send binary files via e-mail in a timely and economical manner, minimizing the effect of the file on other messages flowing through a gateway.

9.5 THUMBSPLUS

In concluding this chapter we will turn our attention to a comprehensive image manipulation program that provides you with the ability to view, locate and organize your images. This program is ThumbsPlus from Cerious Software Inc. of Charlotte, N.C.

Versions

Two versions of ThumbsPlus are provided on the CD-ROM that accompanies this book. The first version, ThumbsPlus Version 2.0e, is a 16-bit version of the program that can operate under Windows 3.1, Windows NT or Windows 95. The second version of the program, ThumbsPlus Version 3, represents the 32-bit version and it can operate under Windows 95 or Windows NT. In this section we will primarily focus our attention on the use of the 32-bit version of Thumbs-Plus; however, we will briefly discuss the installation of both programs.

Installation

Three files are located under the directory THUMBS on the CD-ROM that accompanies this book. The file SETUP.EXE and SETUP.W02 represent the installation files for ThumbsPlus Version 3, and the file THMPLS.EXE represents the installation file for the 16-bit version of the program.

Assuming that drive D contains your CD-ROM, you would select RUN from your Start menu and enter D:\THUMBS\SETUP.EXE if you intend to install the 32-bit version of the program, or

Figure 9.29 The initial ThumbsPlus Welcome! Dialog box displayed during the installation of the 32-bit version of the program

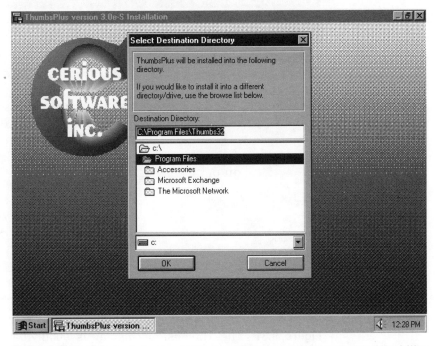

Figure 9.30 The installation of ThumbsPlus provides you with the ability to define the location where the program will be installed

D:\THUMBS\THMPLS.EXE if you want to install the 16-bit version of the program. If you are using Windows 3.1 you would select RUN from the File menu in Program Manager; however, you can only install THMPLS.EXE, as Windows 3.1 is restricted to running 16-bit applications.

Figure 9.29 illustrates the Welcome! dialog box generated by the installation of the 32-bit ThumbsPlus version 3.0e. When you click on the OK button you will obtain the ability to control the location on your hard drive where the program will be installed. By selecting the default location shown in Figure 9.30, the installation process will provide you with the ability to execute the program from the Start button because the entry ThumbsPlus will be placed under the Program entry. Thus, you would then select Start, move your cursor over Programs, select ThumbsPlus, and then select the ThumbsPlus 3.0 icon to execute the program.

During the installation process you will be reminded that ThumbsPlus is a shareware program provided for no charge during a 30-day evaluation period. You can contact Cerious Software on the Web at http://www.cerious.com, via e-mail at support@-cerious.com, via America OnLine at 76352,14, via telephone at

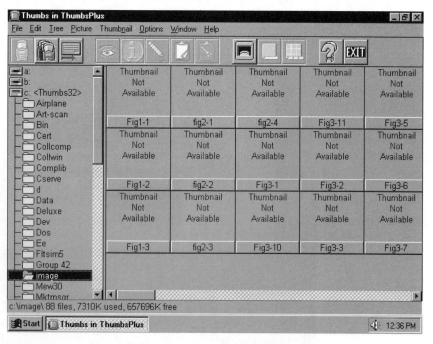

Figure 9.31 The ThumbsPlus display is subdivided into three main areas: a thumbnail viewing area, a directory selection area, and a toolbar of icons that provides shortcuts to frequently used commands

704-529-0200, and via fax at 704-529-0497. You can also use the program to obtain information concerning its registration or write to them at 1515 Mockingbird Lane, Suite 910, Charlotte, N.C. 28209.

Operation

To obtain a brief appreciation for the extensive capability of ThumbsPlus we will examine its use. Figure 9.31 illustrates the program display after the program was executed and this author selected the directory Image. Doing so resulted in the program searching that directory for over 40 types of image file that it supports and listing the filename of each file encountered.

If you examine the program display window you will note that it is divided into three functional areas in addition to the normal Windows title bar and menu bar. The first functional area consists of icons in a horizontal line directly under the Windows menu bar. Those icons represent shortcuts for operations that can also be selected from the menu bar. The first icon on the left, which is not

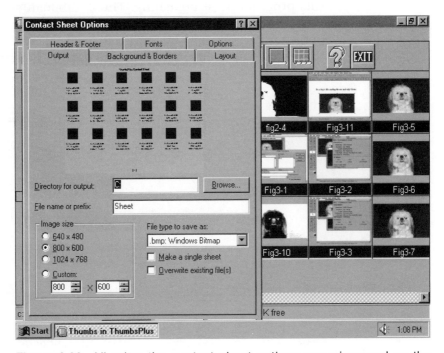

Figure 9.32 Viewing the contact sheet options superimposed on the creation of thumbnails from image files in the Image directory

highlighted at the present time, cannot be selected until thumbnails have been updated. That icon, when selected, represents Make, the double thumb icon to the right of the first icon represents Update, and the third icon from the left that looks like a disk drive with an arrow under it represents Scan drive. Although most of the icons are not highlighted, let us discuss their functions when they are highlighted and how they become highlighted.

If you double-click on the double thumb icon the program will scan each image file in the selected directory and display a thumbnail for each file. Those thumbnails are displayed in the second portion of the display referred to as its viewing area. The viewing area in Figure 9.31 currently lists the file names of the images in the directory that we have selected. The third area of the display is the directory list at the left portion of the display from which we have selected the Image directory on drive C.

Figure 9.32 illustrates a portion of the thumbnail display resulting from double-clicking on the previously mentioned update icon. On the left portion of Figure 9.32 the contact sheet options menu is displayed as an overlay, having been selected from the Thumbnail menu bar. If you examine the contact sheet options you will note that ThumbsPlus gives users a major capability to control how thumbnails are displayed that makes this program most appealing. Thus, although space constraints preclude a detailed examination of the use of this comprehensive program, readers are encouraged to check out its tremendous thumbnail capability, which should easily justify its registration.

INDEX

LOCAL AREA NETWORKING

PROTECTING LAN RESOURCES
A Comprehensive Guide to Securing, Protecting and Rebuilding a Network

With the evolution of distributed computing, security is now a key issue for network users. This comprehensive guide will provide network managers and users with a detailed knowledge of the techniques and tools they can use to secure their data against unauthorised users. Gil Held also provides guidance on how to prevent disasters such as self-corruption of data and computer viruses.
1995 0 471 95407 1

LOCAL AREA NETWORK PERFORMANCE
Issues and Answers
Second Edition

The performance of LANs depends upon a large number of variables, including the access method, the media and cable length, the bridging and the gateway methods. This revised text covers all these variables to enable the reader to select and design equipment for reliability and high performance.
1996 0 471 96926 5

LAN TESTING AND TROUBLESHOOTING
Reliability Tuning Techniques

Network testing is becoming a major requirement in corporate, industry and government computing. This book focuses on networking systems and the testing tools on the market today.
1996 0 471 95880 8

HIGH-SPEED NETWORKING WITH LAN SWITCHES

The demand for switching is on the increase as higher bandwidths are required from LANs, the internet and intranets. This book focuses on different types of LAN switches and how they fit in with current network devices.
1997 0 471 18444 6

VIRTUAL LANs
Construction, Implementation and Management

Virtual LANs allow network administrators to group users in a logical network rather than one based upon physical location. The book examines this new way of setting up networks from an intermediate level.
1997 0 471 17732 6